Climate Change and Sustainable Development

Prospects for Developing Countries

Edited by
Anil Markandya and Kirsten Halsnaes

from Routledge

First published in the UK and USA in 2002 by
Earthscan Publications Ltd

This edition published 2013 by Earthscan

For a full list of publications please contact:

Earthscan
2 Park Square, Milton Park, Abingdon, Oxon OX14 4RN
Simultaneously published in the USA and Canada by Earthscan
711 Third Avenue, New York, NY 10017

Earthscan is an imprint of the Taylor & Francis Group, an informa business

ISBN: 978-1-853-83910-8 (pbk)
ISBN: 978-1-85383-911-5   (hbk)

Typeset by JS Typesetting Ltd, Wellingborough, Northants
Cover design by Danny Gillespie

A catalogue record for this book is available from the British Library

Library of Congress Cataloging-in-Publication Data

Climate change and sustainable development : prospects for developing countries / edited by Anil Markandya and Kirsten Halsnaes.
    p. cm.
Includes bibliographical references and index.
ISBN 1-85383-910-8 (pbk. : alk) – ISBN 1-85383-911-6 (hbk. : alk.)
    1. Climatic changes–Government policy–Developing countries. 2. Sustainable development–Developing countries. I. Markandya, Anil, 1945- II. Halsnaes, Kirsten.

QC981.8.C5 C5113825 2002
363.738'747'091724–dc21

2002006344

Contents

List of Figures, Tables and Boxes

LIST OF FIGURES

LIST OF TABLES

LIST OF BOXES

List of Contributors

Kirsten Halsnaes is a leading economist on climate change studies, covering various developmental, economic and environmental aspects. She has an MSc in economics from the University of Copenhagen and a PhD in economics. Together with Professor Markandya, she was a coordinating lead author for costing methodology issues in the IPCC Third Assessment Report and has also been a lead author on the IPCC Synthesis Report and on their Second Assessment Report. She has contributed to research on environmental economic issues in various areas, including studies of ancillary benefits for the energy sector and transportation, estimation of global markets for carbon reduction measures, and integrated studies of developmental and environmental impacts. Over a period of ten years she has been the principal economist of the UNEP Collaborating Centre on Energy and Environment at Risø National Laboratory, where she holds a position as senior research specialist. Her publications include 'Costing Methodologies' of IPCC, 2001 and 'Mitigation and Market Potential for Kyoto Mechanisms: Estimation of the Global Market Potential for Co-operative Greenhouse Gas Emission Reduction Policies', *Energy Policy*, Vol 30, No 1, January 2002.

Anil Markandya is a leading authority on environmental economics and has worked extensively on climate change issues. He was a coordinating lead author for the climate change mitigation assessment of 2001 and has contributed to research in the areas of climate policy and the assessment of ancillary benefits. He has been professor of quantitative economics at the University of Bath, UK, since 1996 and is currently lead economist at the World Bank on environment and socially sustainable development in Eastern Europe and Central Asia. Over a period of 30 years his previous appointments include academic positions at the University of London, and at universities in the US and Europe, and advisory positions with many government and international organizations. He is co-author of the *Dictionary of Environmental Economics* (Earthscan, 2001), co-editor of *Blueprint 2: Greening the World Economy* (Earthscan, 1991) and co-author of *Blueprint for a Green Economy* (Earthscan, 1989).

Pamela Mason is a researcher in environmental economics in the Department of Economics and International Development at the University of Bath, UK. She has an MSc in economics from the Scottish Doctoral Programme and a PhD in environmental economics from the University of York, UK. Her PhD thesis is on the economic theory of sustainable development and renewable and non-renewable resource use. She has since studied the sustainability of non-renewable resource use and techniques for green national accounting, and teaches environmental economics at the University of Bath. She is co-author of the *Dictionary of Environmental Economics* (Earthscan, 2001).

Anne Olhoff is a researcher in environment and development economics at the UNEP Collaborating Centre on Energy and Environment in Denmark. She has an MSc in economics from the University of Copenhagen and has recently obtained her PhD in economics. Her main fields of study include sustainability and poverty indicators, and distribution and inequality.

Ronaldo Serôa da Motta is professor of environmental economics at the University of Santa Úrsual, Brazil. He received his PhD in economics from University College London, UK. He also holds MScs in environmental economics and environmental planning. As well as having held a number of positions within environment-related governmental and research organizations, he has acted as a consultant to the United Nations Environment Programme and the World Bank on economic instruments in developing countries, and to the Inter-American Development Bank on environmental investment in Brazil. His publications include *Environmental Economics and Policy Making in Developing Countries* (Edward Elgar, 2001) and *Pricing the Earth* (Columbia University Press, 1996).

Leena Srivastava is currently director of the Regulatory Studies and Governance Division of the Tata Energy Research Institute (TERI). She has a PhD in energy economics from the Indian Institute of Science, Bangalore, India. In 19 years at TERI she has worked on a range of issues covering energy, economy modelling, energy pricing studies, environmental economics, climate change issues and, most recently, regulatory reforms. She currently holds the additional position of dean of the Faculty of Policy and Planning at the TERI School of Advanced Studies, which she has held since June 2001. She is a member of several national and international committees and has several publications to her credit.

Tim Taylor is a researcher in the Department of Economics and International Development at the University of Bath, UK. He is co-author of the *Dictionary of Environmental Economics* (Earthscan, 2001) and a contributing author to the Intergovernmental Panel on Climate Change's *Climate Change 2001: Mitigation* (Cambridge University Press, 2001).

List of Acronyms and Abbreviations

ADB	Asian Development Bank
AHP	analytical hierarchical process
ALGAS	Asia least-cost greenhouse gas abatement strategy
C	carbon
CBA	cost-benefit analysis
CDM	Clean Development Mechanism
CEA	cost-effectiveness analysis
CFLs	compact fluorescent lamps
CH_4	methane
CO	carbon monoxide
CO_2	carbon dioxide
Cu	copper
ECU	European currency unit
EKC	environmental Kuznets curve
EU	European Union
FLONA	National Forests (Florestas Nacionais) Brazil
GDP	gross domestic product
GEF	Global Environment Facility
Gg	giga-grams (million grams)
GHG	greenhouse gas
GNP	gross national product
GWP	global warming potential
ha	hectare
HIES	household income and expenditure survey (Botswana)
HIPRE 3+	hierarchical preference analysis
IFC	international finance cooperation
IPCC TAR	Intergovernmental Panel on Climate Change Third Assessment Report
IPCC	Intergovernmental Panel on Climate Change
IUCN	International Union for the Conservation of Nature
IWF	intertemporal welfare function
JI	joint implementation
LPG	liquefied petroleum gas
MCA	multicriteria analysis
MTOE	million tonnes oil equivalent
NGO	non-governmental organization
N_2O	nitrous oxide
NFYR	Ninth Five Year Plan (of India)
NH_3	ammonia
Ni	nickel

NMHC	non-methane hydrocarbon
NO_x	nitrogen oxides
NPV	net present value
ODA	Overseas Development Agency, UK (now Department for International Development)
PM	particulate matter
PM_{10}	PM with a mass median aerodynamic diameter less than 10 micrometers
PMF	progressive massive fibrosis
PPP	purchasing power parity
PV	photovoltaic
PV-maximizing	present value-maximizing
RoSCA	Rotating Savings and Credit Associations
SD	sustainable development
SMS	safe minimum standards
SMU	social marginal utility
SO_2	sulphur dioxide
SRTP	social rate of time preference
SWF	Social Welfare Function
tCO_2	tonne of CO_2
TERI	Tata Energy Research Institute
Tg	terra grams
UNDP	United Nations Development Programme
UNEP	United Nations Environment Programme
UNFCCC	United Nations Framework Convention on Climate Change
VOC	volatile organic compound
VOSL	value of statistical life
WTA	willingness to accept
WTP	willingness to pay
WWF	World Wildlife Fund

Chapter 1

Climate Change and Sustainable Development: An Overview

Anil Markandya and Kirsten Halsnaes

BACKGROUND

Climate change has been universally recognized as a global problem. While, historically, the preponderance of greenhouse gas emissions have been in developed countries, emissions will increase rapidly with expected and needed economic growth in developing countries. Yet, in the years since the constitution of the United Nations Framework Convention on Climate Change (UNFCCC) in 1992, global cooperation on climate change has not developed adequately and the discussion on how to address climate change in the longer term has become polarized. The principal reason for this lack of progress is that in developing countries, climate change is not an important focus of economic or development policy and only recently has it been considered among national environmental policy objectives. Climate change remains too marginal compared to the pressing issues of food security, poverty, natural resource management, energy needs and access, or urban transport or land use to capture the attention of leading actors. Various parties to the UNFCCC, as well as independent scientific analysis, have reiterated that strong and inclusive global cooperation that integrates sustainable development and climate change policy objectives will be needed to address these global environmental issues.

Research on the relationship between climate change policies and those of sustainable development is an emerging international research issue. To date, research in this area has included a broad range of issues, covering development, the social and environmental dimensions of climate change and sustainable development. The focus particularly has been on how to facilitate a global climate change policy that both meets the objectives of sustainable development and, at the same time, is based on a broad consensus between stakeholders in industrialized countries, countries with economies in transition and developing countries.

The focus of this book is on a subset of these issues – namely, on the relationship between general sustainable development policies and climate change strategies in developing countries. In this context sustainable development policies are important in determining future greenhouse gas (GHG) emissions and the potential and related costs of climate change mitigation policies. At the same time, climate change policies have a number of impacts on sustainable development policy objectives.

Current international climate change policies have been uniquely driven by global environmental policy concerns, and very little attention has been given to local development and the environmental impacts of specific policies. However, from the local perspective, ancillary benefits of climate change policies, such as increased energy efficiency and the health impacts of local air pollution, may be significant and may therefore be very important in promoting local action. These local policy impacts, however, have not so far been addressed appropriately in climate change policy analysis. This is in contrast to the evidence from many ongoing national development programmes that demonstrates that energy efficiency improvements and other climate favouring activities emerge as side-benefits of sound development programmes. Price reforms, agricultural soil protection, sustainable forestry, energy sector restructuring – all undertaken without any reference to climate change mitigation or adaptation – have had substantial effects on the growth rates of GHG emissions. This observation suggests that it may often be possible to build environmental and climate policy upon development priorities that are vitally important to decision-makers. Furthermore, it opens the perspective that climate change policies may be seen not as a burden to be avoided but as a potential side-benefit of sound and internationally supported development strategies.

As mentioned above, sustainable development is a broad concept that encompasses a wide range of issues relating to development, equity and environmental policy aspects. The international literature includes several hundred alternative definitions of the concept of sustainable development. The aim of this book is not to be comprehensive in addressing the various aspects and understandings of sustainable development, but rather to contribute to the development of a methodological framework for and a practical approach to assessing the sustainable development impacts of GHG emission-reduction projects. Some of the key methodological issues involved in developing a framework for assessing the sustainable development impacts of GHG emission-reduction policies are outlined and discussed. The focus is primarily on the development of a practical basis for assessing projects in developing countries. The book addresses a number of issues that are relevant to the establishment of an international policy regime, including the issues covered by the UNFCCC that facilitate the implementation of policies in developing countries that both reduce GHG emissions and support sustainable development policy objectives.

The core analysis in this book is based on economic methods, but links to sociology, ethics and technology assessments are established. There is an extended debate about the methods of analysis that are most suitable for studying sustainability, and generally there is a recognition that the different disciplines cannot stand alone. Broadly speaking, economics contributes to the evaluation of the efficiency of different mitigation policies, and to the careful assessment of the costs of the policies, while the sustainability concept addresses a number of issues that are more closely linked to, for example, moral and political judgements, in particular intra- and intergenerational equity. Sociological methods emerge as important in dealing with institutional issues, and multi-criteria assessment approaches help us to understand how to rank options when several objectives are relevant and when uncertainty plays a major part.

As noted above, the decision-making approach applied here is primarily a technical paradigm inspired by economics but structured to have an interface with broader decision-making approaches. Accordingly, some sort of decision-making process is presumed to exist for the selection and valuation of national priorities in the context of sustainable development objectives.[1] The methods suggested in this book are intended to support the political decision-making process largely by ensuring that it results in choices that have some internal consistency. It is on this background that sustainability indicators are specified and assessed in a way that is consistent with evaluation approaches like cost-effectiveness analysis (CEA), cost-benefit analysis (CBA) and multicriteria analysis (MCA).[2] The use of such approaches for integrated assessment of various policy impacts is discussed, among others, by Kirkpatrick and Lee (1998 and 2000).

The point of departure for this book is the existing literature on sustainable development, much of which, it is widely recognized, is rather conceptual and 'macro' in its orientation. It has been developed to address global intergenerational resource allocation and equity aspects that are very general in nature. Hence it offers little in the way of advice at the project or programme level, except in a rather ad hoc way, and there is a 'gap' between the general scope of the literature and the specific issues that are relevant in relation to small-scale policy efforts like GHG emission-reduction projects. The sustainable development consequences of small-scale policy efforts, like projects, are difficult to track and to separate from other factors that are influencing the larger systems. The approach taken by the authors, therefore, is to suggest a number of sustainable development indicators of projects and programmes and to look at the implications of using this information as input for policy evaluation. The suggested indicators are not intended to be comprehensive but can be considered as useful 'pointers' to the sustainable development implications of policies.

Finally, it is important to recognize that an integrated assessment of the economic, environmental and social sustainability dimensions of policies is very complicated and resource-demanding. The recommendation is thus to balance methodological sophistication, availability of analytical skills and data, and the strong requirement to facilitate a transparent analysis and decision-making process. In this way a relatively simple initial approach is recommended, where a small number of impact indicators are assessed based on a well-established methodological approach.

STRUCTURE OF THE BOOK

The book is structured as follows. Chapter 2 provides an introduction to the general literature on sustainable development, including its conceptual foundations, and looks at its implications for climate change mitigation and how it can be applied to the climate change debate, especially at the project level. Chapter

3 starts from a review of the literature on sustainable development *and* climate change, which, interestingly, has rather different roots from the theoretical literature reviewed in Chapter 2. It also reports on some of the measured impacts of climate change mitigation projects that reflect their broader sustainable development contributions and draws some conclusions. Chapter 4 is devoted to addressing the social capital aspects of climate change projects and assessing how they can be measured in a way that is consistent with the other dimensions. Chapter 5 reviews different rules for the formal ranking of climate change mitigation projects and reports on their application in a case study for Botswana. These rules are applied in a context that incorporates at least some of the sustainability indicators of interest. Chapter 6 focuses on the equity dimension; the different ways in which it arises in the sustainability context and how it can be made operational. Some numerical calculations are reported for the same case study as in Chapter 5 – Botswana. Chapter 7 reports on the extended mitigation cost analysis that is applied to four countries – Zimbabwe, Botswana, Mauritius and Thailand. Finally, Chapter 8 provides two more case studies carried out using different methods of assessment. The first is from Brazil, which uses a more judgemental approach and the second is from India, which uses multicriteria analysis.

CHAPTER 2: A CONCEPTUAL FRAMEWORK FOR ANALYSING CLIMATE CHANGE IN THE CONTEXT OF SUSTAINABLE DEVELOPMENT

This chapter provides a discussion of the conceptual foundations of sustainable development and then derives some local level indicators of sustainability. The literature defines three dimensions to sustainability – economic, environmental and social. Each of these can be thought of as being 'capital' stocks, which can be used sustainably – ie, in such a way that overall stocks are not allowed to decline; or can be used in an unsustainable way, which means that the stocks are allowed to deteriorate or to be consumed over time.

Indicators exist at the national level that attempt to measure the economic dimension of sustainability based on this notion. These indicators are classified as reflecting weak sustainability or strong sustainability. The difference between the two is that weak sustainability assumes that different forms of capital can be substituted along a sustainable path. Strong sustainability, on the other hand, assumes that substitution between, for example, man-made capital and natural capital is very limited. The chapter discusses the rules and indicators that have been developed in the literature for both weak and strong sustainability.

Applying these concepts at a local level is possible but requires the measures to be made operational. The chapter describes how the concepts of strong and weak sustainability can indeed be made operational in the context of a project, and from this a wide range of indicators is developed, most of which are in the literature on sustainability at the local level, but some of which are not.

The discussion of indicators is organized around the following criteria: data availability, theoretical rationale, presence of a time dimension and relevance to projects. As a general point, specific indicators will apply in specific cases and we cannot offer generic advice for all projects. Nevertheless, some general conclusions can be drawn:

1   The environmental indicators of greatest value are:
    • changes in water and air pollutants accompanying the GHG mitigation project;
    • changes in the use of raw materials and minerals accompanying the same;
    • changes in sustainable or unsustainable forestry practices induced by the projects;
    • an important indirect indicator is the impact on the use of renewable energy in the country.
2   The time dimension is important in most of the key indicators.
3   On the socioeconomic indicators, the most important direct impacts are those on employment and food supply (in sequestration projects). The most relevant indirect impacts are on productivity and regional value added. A useful innovative indicator is the contribution of the project to genuine savings, although this has yet to be implemented through empirical studies.
4   The indicators of social acceptability and social capital are not *indicators* in the usual sense of the term because they are difficult to link to individual projects. Their main usefulness is as a guide to appropriate prior 'operating conditions' for the projects. Ideally, we should ensure that these conditions (for example, a management plan, high social mobility, good horizontal networks, etc) are present. The question then arises, what if they are not? Projects undertaken in the absence of what we will call 'good reference operating conditions' will need further support if they are to be successful. This issue is raised again later in the book.

The indicators discussed in the chapter cover a wide range of impacts and it is not expected that all projects will use all of them. One of the areas of further work is to provide more guidance on where particular indicators should be used. The discussion provides, however, a useful 'shopping list' of indicators, based on the most recent literature on the subject.

CHAPTER 3: A REVIEW OF THE LITERATURE ON CLIMATE CHANGE AND SUSTAINABLE DEVELOPMENT

The application of the sustainable development concept to climate change studies has thrown up a number of theoretical and practical complexities. These are related to the very broad policy agenda that sustainable development is trying to address and to the very ambitious scope of the work. This scope is formulated by the Intergovernmental Panel on Climate Change Third Assessment Report (IPCC TAR) (Banuri et al, 2001) as enabling the assessment of the synergies and trade-offs involved in the pursuit of multiple goals – ie, environmental conservation, social equity, economic growth and poverty eradication. The review in this

chapter does not identify any studies that address the broad range of sustainability policy objectives and how they can be integrated in climate change
mitigation studies, but it identifies a number of studies that make a first attempt
at integrating a number of important indicators. These include ancillary benefit
studies and studies that have assessed climate change mitigation projects in
relation to selected sustainability impact indicators representing development,
environmental, social and in some cases technological implementation dimensions of sustainable development.

The ancillary benefit studies have focused on short-term side-impacts of
climate change mitigation policies, in particular local air and water quality
impacts. The sustainability impact studies have also included these local environmental impacts, but have suggested a number of additional indicators to reflect
development (such as employment generation) and social impacts (such as
income distribution). The approach of ancillary benefit studies has typically been
a CBA or CEA, while the sustainable development impact studies have been
based on various approaches, including MCA with quantitative and qualitative
information and CBA.

The review of empirical studies identifies many similarities in analytical
approaches across papers. Many of them have demonstrated that there is a
potentially wide range of climate change mitigation policies, which have significant co-benefits on the local environment and on development. These benefits
can be very significant in a number of developing countries, especially those
characterized by high local pollution levels and unemployed labour. Some climate
change mitigation projects might also have a number of negative side-impacts on
the local environment and employment. A strong conclusion across all the studies
is that the impacts of climate change mitigation projects on sustainable development are very site and context specific, and the actual impact should therefore be
assessed from case to case.

A number of methodological issues that need more work are also highlighted. These include the need to build closer links between the measured
impacts and the theoretical dimensions of sustainable development with its three
types of 'capital'. Another need is to develop a methodological framework that
facilitates conceptual and analytical consistency across studies. The main components of such a framework would be the consistent definition of cost concepts
and sustainable development impact indicators, baseline scenario approaches and
assumptions, terminology and format of qualitative and quantitative information, and classification of analytical approaches.

CHAPTER 4: ASSESSING SOCIAL CAPITAL ASPECTS OF CLIMATE CHANGE PROJECTS

The social dimension of climate change reduction is probably the most difficult
to analyse in a satisfactory manner and yet it is one of the most important. Two
aspects of social capital are considered in this chapter: linkages *between* and
within groups, and the interaction between *state and society*. Focusing on the
nature of relationships between and within groups and between the state and

society, and drawing on insights from economic theory, a simple analytical framework for the assessment of social capacity is developed. The main conclusions to be drawn from the analytical framework are that:

- The selected options, the implementation policies and the mechanisms for implementation must reflect the prevailing social capacity if implementation is to be successful.
- The social capacity has implications for the implementation of different options. Some types of projects and strategies can only be implemented if the social capacity is strong.
- Implementation of different options will in most cases necessitate social capacity-building activities. The relevant social capacity-building activities will depend on the social capacity in a given situation.

The conventional approach to barriers and barrier removal suffers from a weakness because it does not facilitate an assessment of whether the barriers are interrelated and rooted in the social organization of society. The basic problem is that a coherent framework for assessing the overall institutional structure and functioning of the economy is missing. The inclusion of social capital aspects permits the development of such a framework and provides a link between the implementation of GHG emission reduction options and the concept of social capital.

Social capacity-building involves activities that enhance the capacity of individuals, institutions and society to undertake an activity or project in accordance with general economic policies and development programmes. In this way capacity-building efforts can become catalysts for the market development and commercialization of the specific options under consideration. As a consequence, social capacity-building efforts are assumed to be temporary activities that establish a permanent capacity for a given option. The chapter outlines how the costs of such activities may be measured in the context of GHG emission reduction projects.

Implementation costs are one of the most critical components in providing more reliable estimates of mitigation potential and related policies. It is not easy, however, to develop a framework for assessing implementation costs that addresses the many specific national issues that will be critical in strategy development. Some potentially useful indicators are presented, and examples of their inclusion in project assessment are given. To facilitate comparison with the other chapters, the actual examples are based on the same case study as used for the earlier chapters – Botswana. In that case it turns out that including an estimate of the implementation costs does not change the rankings of options, but this may not always be the case. Indeed, ignoring implementation costs may result in misleading policy information, with projects that have high true costs being selected and then found to fail when the implementation component is not adequately accounted for. Furthermore, ignoring such costs may result in projects being selected that have the least significant development impacts, because high development benefits can go with high implementation costs.

CHAPTER 5: ANALYTICAL APPROACHES FOR DECISION-MAKING, SUSTAINABLE DEVELOPMENT AND GHG EMISSION-REDUCTION POLICIES

Chapter 5 considers how sustainable development concerns can be addressed in an evaluation of policy options that meet global objectives in the form GHG emission reductions and local development objectives, including economic, environmental and social goals. The preferred approach is to use a public planning framework, where technical assessments of a range of policy objectives are undertaken, based on priorities that are established in an interactive process with policy-makers and other stakeholders, or on the basis of stated preferences in national development programmes. Public planning in this context is understood as a multi-stage process, which includes separate steps that identify policy goals, screen projects and policies, and transform goals into objectives that can be assessed in technical assessments, leading to an assessment of policy impacts which finally are used to generate information to the policy-making process.

The structure and critical assumptions included in the alternative technical approaches of CBA, CEA and MCA are compared. These approaches exhibit a number of structural similarities. They are all based on an objective function that includes specifically chosen policy priorities (and excludes other possible priorities). In this way they have a normative character when applied to public planning problems. Another structural similarity is that they all seek to integrate different policy objectives into a more general decision-making framework, which in some cases will imply the use of weights to value individual objectives. It is suggested, as far as possible, to limit the number of trade-offs that are considered in the technical assessments through systematic analysis of the 'efficiency' of different options (which means excluding those that are inferior on all objectives) and through introducing 'safe minimum' standards for given impact areas.

A major difference between the welfare economic-based approaches, such as the CBA, CEA and the MCA is the procedure that is used to establish weights for different policy objectives. The economic approaches primarily establish these weights on the basis of stated preferences of individuals in markets or in constructed markets, while the MCA tries to establish weights that reflect the relative performance of the policies and projects, which can be derived through a dialogue with policy-makers.

The differences and similarities between the different technical approaches were tested in an evaluation of five potential GHG emission reduction options for Botswana. The analysis was performed for:

1   CBA based on the net benefits;
2   CBA based on the cost-benefit cost ratio;
3   CEA based on gross financial costs of the project per unit of GHG removed;
4   CEA based on the net social costs;
5   MCA based on equal weights for financial cost, value of employment and emissions reductions (with each objective measured on a normalized scale of zero to one); and

6   MCA based on weights for the above-mentioned objectives being altered to give greater importance to reductions of local air pollutants.

The project rankings that emerge from the two different versions of, respectively, the CBA, the CEA and the MCA lead to different project rankings. This result highlights the importance of the specific weighting of individual policy impacts as well as of the implicit political criteria on the results.

One conclusion drawn from this comparative study is the crucial role that public planning exercises like the one considered here play in separating the initial selection of general policy priorities from the subsequent evaluation of the performance of different options. This separation is essential in order to avoid the technical assessment becoming merely a sophisticated tool for the confirmation of pre-established policy priorities.

Another conclusion that can be drawn from the comparative assessment of the technical approaches is the importance of transparency in the reporting of the data inputs, the objectives and the objective weights. In the same way, final conclusions should be discussed critically in relation to the uncertainties inherent in such assessments, in particular with regard to the inclusion of impacts that are beyond the scope of conventionally marketed goods and services. Following this, it is recommended that sensitivity analysis of key parameters be conducted and other potential assumptions that could lead to other conclusions be discussed.

Finally, we should note that the weights and the choice between the technical approaches themselves may depend on whose perspective we take. For a developing country the weights will differ from those used by a donor or an industrialized country. Furthermore, if the developing country has no target, it may prefer to select its priorities based on CBA or MCA, rather than the CEA approach.

CHAPTER 6: SUSTAINABILITY IN CLIMATE MITIGATION: INTEGRATING EQUITY INTO PROJECT ANALYSIS

Equity considerations are of importance in determining the sustainability and potential for the implementation of climate policy. Intergenerational equity concerns highlight the necessity for action, but once action has been deemed necessary intragenerational concerns are of greater importance. With increased focus on issues such as poverty alleviation and other development actions as co-benefits of climate mitigation strategy, the determination of the distributional implications of different policy measures takes on a greater degree of importance.

The chapter advocates that one begins by identifying groups of individuals who will be affected by the projects and ensuring that adequate data are collected to assess the income and other changes in their well-being resulting from the project. The process by which these groups are identified and the analytical steps in arriving at a ranking that includes equity considerations is discussed at some length.

One method of integrating equity considerations into the standard analysis is that of using distributional weights. The selection of the weights to be applied is, of course, important and should be determined from the priorities assigned to the equity issue by policy-makers. This chapter does not advocate that this technique is without flaws, but the use of weights does allow easy comparison of projects and their impacts. The technique is applied to the ranking of the same projects that were evaluated in Chapters 4 and 5 – from the case study for Botswana. Different weights are used, with both consumption and income as the basis for the calculation. While it does not make much difference to the ranking whether a consumption or an income base is used, it does make a difference to how the project is funded (whether it is from taxation or self-funded, from charges on GHG emitting goods). It also makes some difference as to what weight is given to inequality relative to the standard benefits – ie, how much more value is attached to a dollar going to a poor person as opposed to a rich person. The overall conclusion is that equity really matters and can change both the formal rankings as well as the selection of projects from a political standpoint significantly.

The case study at the end of the chapter illustrates how distributional weighting of the control cost and ancillary benefits of climate mitigation projects may impact project analysis. As knowledge on the distribution of ancillary benefits expands, such weighting may yet further impact on the analysis of projects.

CHAPTER 7: CASE STUDIES FOR ZIMBABWE, BOTSWANA, MAURITIUS AND THAILAND

In Chapter 7 a more detailed evaluation of GHG mitigation options is reported. For each country a range of options is evaluated, looking at the financial costs as well as the costs and benefits in terms of changes in other fossil fuel emissions and changes in employment. The methods by which the latter two can be measured in monetary terms are also discussed. The data are from international study programmes in the individual countries, supported by United Nations Environment Programme (UNEP), the Asian Development Bank (ADB), The Global Environment Facility (GEF) and United Nations Development Programme (UNDP).

As the actual estimates of the wider benefits are very approximate, the results are largely illustrative of the method. Indeed, one of the recommendations is that primary work is needed in developing countries to arrive at more accurate estimates of the costs and benefits arising from these secondary impacts of the mitigation programmes.

The financial and social cost assessment is based on a cost-effectiveness approach. An assessment of the financial and social costs of GHG emission-reduction projects have been considered to provide very important information in relation to project evaluation currently discussed in the context of international climate change finance and the UNFCCC that clearly states that climate change policies should meet sustainable development objectives.

The social cost concepts defined and applied in this project assessment are an example of how a number of local development impacts of GHG emission-reduction projects can be assessed. It can very well be argued that the sustainable development impacts of such projects will range beyond the actual employment, local air pollution and coal-mining impact indicators included in this study. An assessment of GHG emission-reduction projects might therefore include a wider range of indicators that can be assessed quantitatively as well as qualitatively. Despite the limited number of indicators assessed for the GHG emission-reduction projects considered in this study, it is worth noticing that the magnitude of the GHG emission-reduction costs as well as the cost-effectiveness ranking of projects are closely related to the financial and social cost-accounting framework. A gross financial cost perspective that primarily reflects the capital costs of the options per unit of GHG emission reduction results in a cost-effectiveness ranking of the projects that is significantly different from the ranking that comes out of a net financial cost-based and a social cost-based ranking. If the gross financial costs are taken as the perspective of project donors, the actual projects initiated from that perspective will not be the same as the ones that would emerge from a greater emphasis on local benefits in the form of energy savings and social improvements.

Another general conclusion is that in all cases, the net financial costs as well as the social costs are lower than the gross financial costs. This implies that a project host country receives local benefits if an international donor is willing to supply the gross financial cost of implementing the project. The actual magnitude of the benefits in the form of energy savings and social benefits are, however, very site specific. The energy savings depend on the efficiency of the baseline energy system, while the social benefits depend on a broad range of site-specific and time-specific issues such as employment, local air pollution and health related issues. In particular, the net financial costs and the social costs will, for similar categories of GHG emission-reduction projects, vary from case to case.

CHAPTER 8: THE CDM AND SUSTAINABLE DEVELOPMENT: CASE STUDIES FROM BRAZIL AND INDIA

Chapter 8 reports on two studies, one from Brazil and one from India, that carry out a similar analysis but from slightly different perspectives. The Brazil study looks at two sequestration options and a number of energy efficiency options from two perspectives – ie, private profitability and secondary benefits. It does not try to rank them using an MCA or other method but merely reports on the complex picture, taking the view that policy-makers would prefer to receive information in this form.

The private profitability analysis leads to the following:

1 Sequestration options: If a market for Clean Development Mechanism (CDM) projects were to exist, private investors seeking CDM rents would be more willing to undertake plantation projects than native forest management, since the former offers higher profitability against lower learning costs, reasonable scale effects and, above all, lower carbon break-even prices. CDM buyers would also go for plantation due to low leakage rates.
2 Energy efficiency options: Industrial cogeneration of electricity has by far the highest private return. Biomass electricity and wind energy options are at the same profitability level as plantation. Ethanol production is, however, not

privately profitable at all and it would only be a CDM option if it is under-taken with government intervention.

However, when secondary benefits are taken into account, a different picture emerges. More specifically, the analysis of secondary benefits points to the following:

1 Environmental benefits:
   • Native forest management options, particularly concession forests, offer a great deal of secondary benefits of great relevance for biodiversity protection.
   • Biomass electricity as charcoal substitution can also assure air pollution benefits.
   • Wind energy offers more air pollution benefits while biomass is more prone for biodiversity protection.
2 Development and equity impacts:
   • Plantations are more important for the activity level of the economy as a whole but less for the regional economy, although they can affect the trade balance deficit negatively. In terms of regional benefits, private sustainable logging in native forests is more relevant. Ethanol and biomass electricity, on the other hand, capture most of all development gains.
   • Equity issues are in favour of native forest management when they affect low-income classes at the project's output, costs and ecological benefits, although they generate more negative impacts from displacement activities than plantations.
   • Ethanol and biomass electricity combine development gains with equity ones.

Thus the partial qualitative analysis presents a mixed picture of which option is the 'best'. What is clear, however, is that private profitability, by itself, has no definitive link pattern to secondary benefits – ie, market forces alone will not be able to select CDM options which have, at the same time, high private cost-effectiveness and high positive linkages to ecological and social benefits.

For India, the study looks at 22 energy-related projects in the context of the development priorities of the country as articulated in the Ninth Five Year Plan (NFYP). These include, inter alia, economic growth (with greater labour absorption) and environmental sustainability factors, which are reflected in the criteria chosen for ranking the chosen options. The options were selected based on their consistency with national development priorities; the relatively high level of energy consumption in the base activity; and the relatively large GHG reduction potential offered by the abatement technology.

The study notes that, which of the above options prove best depends on one's perspective. Developed country investors will be concerned primarily with a project's abatement cost, financial risk and feasibility, while India will be more concerned with the development and environmental benefits that arise, and their consistency with national priorities. Different parties to a transaction will weigh project characteristics differently, suggesting a priori that developed and developing country participants would rank overall projects differently.

To explore how projects meet national priorities and which, if any, would satisfy all parties, projects are assessed both on the basis of their development benefits from the perspective of the NFYP and their carbon abatement cost. Projects are additionally evaluated against their feasibility and other environmental benefits.

Interestingly, for three of the four sectors for which comparisons can be made, the two highest ranked options based on the cost of carbon offset are also the two highest ranked options based on their co-benefits. This suggests that there is a high degree of overlap between those projects that would be prioritized by carbon-focused investors and those that are in the best interest of India. Only the renewables for power generation category shows a mismatch in the first-choice as perceived by the market criteria alone and the wider criteria that make up the AHP scores. Also, in conventional power generation and renewables for agriculture, some of the lower rankings are different.

For the abatement options reviewed, there also seems to be a high degree of overlap between those projects that are available at lowest cost and those which are most consistent with national priorities and offer most to India in terms of development and environment co-benefits.

The study notes that there are improvements that could be made in the analysis. One would be by extending it to compare all projects together, rather than merely within sectors. This would allow policy-makers to know which sectors should be emphasized before establishing which projects within a sector make most sense. This will require the weighing of each sector to project relative importance of individual sectors to overall economic growth. From the investors' standpoint, the total potential in each sector and the transactions costs is an important factor. These factors will play a role only if options are compared across sectors and not in a study that looks at sector priorities. Finally, more details on the projects themselves would allow for more accurate evaluation. For example, the health benefits of cleaner coal technologies are sensitive to the precise location of the plant to be modified.

CONCLUSIONS

The assessment of climate change mitigation options in developing countries is a subject of considerable interest in the context of the UNFCCC and international cooperation on climate change finance. At one level it is important to know the direct financial costs involved, the amount of carbon emissions reduced and hence the cost per tonne of carbon. The selection should then be based on whatever turns out to have the lowest cost per tonne of carbon. Unfortunately, however, matters are not that simple. While any outside investor may only be interested in such gross financial costs, the country itself will have wider considerations that it wishes to take into account. Precisely what these are will vary from country to country, but some common factors will apply widely. Among them are the impacts of the project(s) on local air quality and other environmental indicators, employment, equity and sustainable development in general.

This book examines how these wider considerations can be incorporated into the process of evaluating the options. It reviews the literature on sustainable development and offers some indicators that can be used for the analysis of sustainability at the project level. It also reviews the existing literature on integrating sustainability in climate change mitigation analysis and makes some recommendations on how effective present methods are and how they can be improved. Among the different elements of a sustainability assessment, the most neglected in the present work are the social and developmental aspects, including the impacts of projects on social networks and their dependence on them for success in implementation. This study suggests some ways in which these issues can be taken into account. Another important dimension is that of equity, and careful consideration of equity factors can change the policy priority ranking of different options.

The formal methods used for the evaluation of options are reviewed in the book and case studies of their application are presented. Among the formal ones, traditional CBA and CEA can be modified to pick up many of the sustainability concerns, and this book shows how this can be done. At the end of the day the choices are, of course, to a large extent political, but there is a case to be made for presenting decision-makers with a procedure that simplifies the complex information into a single measure. But there are some who argue against that and suggest a more detailed presentation of the information, such as that made for the Brazil case study. The authors believe that there is a need for both. The discussion of the issues and methods of presentation provided in this book should contribute substantially to the effective analysis and presentation of the relevant information to policy-makers.

NOTES

1   The decision-making process can take various forms and can be based both on official national or sectoral development plans or on specific discussions with various national stakeholders. Technically speaking, it implies that a sort of decision-making approach can be used to specify the arguments in the objective function of the assessment and to specify weights for these arguments.
2   This can be seen as a contrast to broader decision-making approaches that emphasize stakeholder participation and give less weight to internal methodological consistency. An example of such an approach is Social Impact Assessment (Kirtpatrick and Lee, 1998).

REFERENCES

Banuri, T, Weyant, J, Akumu, G, Najam, A, Pinguelli, R, Rayner, S, Sachs, W, Sharma, R, Yohe, G (2001) 'Scope of the Report: Setting the Stage: Climate Change and Sustainable Development' in B Metz, O Davidson, R Swart and J Pan, *Climate Change 2001: Mitigation – Contribution of Working Group III to the Third Assessment Report of the Intergovernmental Panel on Climate Change*, Cambridge University Press, pp73–114
Kirkpatrick, C and Lee, N (eds) (1998) *Sustainable Development in a Developing World: Integrating Socio-economic Appraisal and Environmental Assessment*, Edward Elgar, Cheltenham
Kirkpatrick, C and Lee, N (eds) (2000) *Sustainable Development and Integrated Appraisal in a Developing World*, Edward Elgar, Cheltenham
UNFCCC (1997) *Kyoto Protocol to the United Nations Framework Convention on Climate Change (UNFCCC)*, FCCC/CP/1997/L.7/Add.1, Bonn

Chapter 2

A Conceptual Framework for Analysing Climate Change in the Context of Sustainable Development

Anil Markandya, Kirsten Halsnaes, Pamela Mason and Anne Olhoff

INTRODUCTION

As we noted in the overview, the term sustainable development is understood with wide differences in the development and climate change communities. As sustainable development is a key concept in the climate change debate, it is important to understand the different strands of thinking that have gone into, and are ongoing in, the sustainable development paradigm. This chapter serves to provide a guide to that literature and to interpret it in the context of the climate change discussions, notably those related to strategies for the mitigation of greenhouse gases (GHGs).

This chapter provides an introduction to the sustainable development literature. This can be a major undertaking, given the extremely wide range of contributions on the topic. We undertake the task because we believe it will clear up some of the thinking on the subject and provide a useful background to those who will be involved in assessing the sustainability dimension of climate change mitigation policies. The section includes a conceptual discussion about economic, social and environmental sustainable development aspects and suggests a framework for how sustainable development can be understood in relation to GHG emission reduction projects. At times it may seem abstract and unrelated to the real world issues that are the focus of this book, but we urge patience on the part of the reader. It is from these conceptual foundations that a number of alternative rules for evaluating sustainable development policies are derived. We go on to examine how such policies can be integrated into an analysis of GHG emission reduction policies. In this context it is important to appreciate the lack of integration at the conceptual level between measures of sustainability that focus on the different dimensions – the economic, the environmental and the social.

Measures for sustainability, including rules for strong and weak sustainability, are introduced, and these rules are linked to project appraisal. A number of specific indicators that can be used to reflect economic, environmental and social sustainability impacts of projects are suggested. The latter is a very practical issue and the highly theoretical discussions have to be brought down to earth before they can be applied. The final section underscores an important aim of this chapter, which is to suggest a set of indicators that reflect economic, social and environmental aspects of sustainable development to be assessed in relation to GHG emission-reduction projects. Several conceptual issues are involved in this exercise, including the establishment of an operational sustainable development definition applied to GHG emission-reduction projects, the selection of indicators that represent important aspects of this sustainable development concept, and finally the evaluation of the sustainable development impacts identified for specific projects. The scope of the current chapter is to consider the conceptual part of developing operational sustainable development indicators, while the evaluation of indicators for specific projects falls under the scope of other chapters of this book.

SUSTAINABLE DEVELOPMENT: THE BACKGROUND TO THE DEBATE

The Earliest Definition of Sustainable Development

The term 'sustainable development' has its origins in the International Union for the Conservation of Nature's (IUCN's) 1980 World Conservation Strategy report (IUCN, WWF and UNEP, 1980), but it was with the World Commission on Environment and Development report, entitled *Our Common Future* (1987) that the term gained broad currency.[1] The commission defined sustainable development as 'development that meets the needs of the present without compromising the ability of future generations to meet their own needs'. This definition, while useful in drawing attention to concern with the long-term implications of present-day development, asks as many questions as it answers. What constitutes 'needs', and how will these change over time? What reductions in the options available to future generations are acceptable and what are not? The operational aspects of sustainable development were not answered by the Brundtland Commission, although the report itself gave strong hints that the environmental degradation resulting from today's economic policies was a major source of concern from a sustainability viewpoint.

The first attempts to make the concept more precise were theoretical rather than practical. They focused on the economic and the environmental dimensions of the debate. From the economic perspective, some of the earlier contributions (Pearce, Barbier and Markandya, 1990) suggested that sustainable development should imply that *no generation* in the future would be worse off than the present generation. In other words, society should not allow welfare to fall over time.

The Economic Dimension of Sustainable Development

Optimal Growth Models and their Sustainability Implications

The more theoretical literature developed the issue as an extension of the 'optimal growth' literature described in this section. If a society was to maximize the welfare of all generations, present and future, how should it allocate resources over time? The models that have been used for this analysis are sophisticated and complex, and it is not appropriate to go into details about them here. However, one or two aspects about them are worth noting. First, the models maximize discounted welfare. This means that the welfare of future generations is given a lower weight than that of present generations. Why is that? Again, the reasons are complex but essentially boil down to saying that people attach progressively less value to utility as it moves further into the future. They do so because the future is more uncertain and because future generations will have access to more technology and capital than present generations. This preference for present versus future utility is captured through the discount rate, and the earliest models were those of Cass (1966), Koopmans (1965) and Malinvaud (1965). These 'neoclassical' models identify the optimal growth path by maximizing an intertemporal welfare function (IWF), which consists of the present discounted value of all generations' utilities. The discount rate measures the rate at which utility becomes less important as it moves further away in time. The social planner derives the optimal solution by maximizing the IWF subject to the budget constraint:

$$\text{Investment} = \text{Output} - \text{Consumption}$$

The budget constraint means that the cost of postponing consumption by investing in capital is offset by the return on capital, which permits higher consumption in the future. This type of model is called a present value (PV)-maximizing model. The model yields the solution that, given the initial level of the capital stock, the economy must invest up to the point at which the marginal product of capital, or the interest rate, is equal to the discount rate. Therefore, if the initial capital stock is relatively low, investment will be positive and consumption will rise towards the equilibrium level at which the interest rate is equal to the discount rate – ie, $F_K = \delta$.

This solution is illustrated in Figure 2.1. In the figure, the initial capital stock is K^0 and the initial consumption level is C^0. Here, the cost of delaying consumption is more than offset by the productivity of the investment, so it is worthwhile to invest. Consumption rises and investment remains positive until the equilibrium capital stock, K^*, is reached.

If, on the other hand, the initial capital stock is so high that the interest rate is lower than the rate of discount, then investment will be negative and consumption will fall towards the equilibrium level.

This model illustrates the relationship between discounting and intertemporal well-being, since it shows that the optimal equilibrium level of consumption depends on the utility discount rate. Moreover, it illustrates the relationship between discounting and sustainability, since if the initial capital stock is greater than the equilibrium level, the optimal solution is to *consume* some of the initial

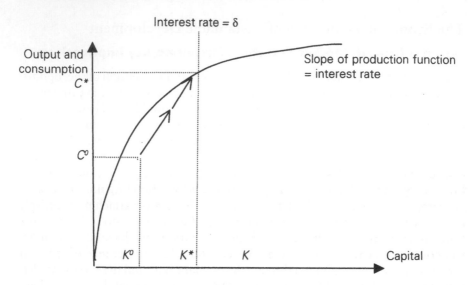

Figure 2.1 *The optimal path of consumption with a constant utility discount rate*

capital stock. In this case consumption falls over time and so the path is 'unsustainable', if sustainability is interpreted as non-declining welfare.

In this model the only asset is man-made capital. This is accumulated over time from low levels, so that the PV-maximizing criterion tends to lead to a sustainable consumption path. In particular, as technical progress increases the productivity of capital, this increases the incentive to invest and increase the capital stock. However, if the capital stock were inherited by the economy rather than being built up by it, it is possible that the initial stock could be higher than the equilibrium level, and in this case the optimal path would be unsustainable.

In reality, however, man-made capital is not the only asset relevant to welfare. One should also account for environmental capital, such as stocks of natural resources. Since these tend to be inherited by economies rather than being built up, the initial stock may well be higher than the final stock. If the neoclassical model described above were extended to such capital, it would imply that the marginal product of environmental capital was less than the discount rate, and the optimal solution would be to consume the stock unsustainably. This illustrates one problem of pursuing the neoclassical solution to economic policy when environmental assets are brought into the picture.

The situation becomes even more problematic when man-made and natural capital stocks are considered in the same framework. This was first analysed by Dasgupta and Heal (1974) in their seminal paper on the efficient use of non-renewable resources according to the PV-maximizing criterion. They set up the optimization problem to maximize the PV of the utility path with a constant discount rate. They demonstrated that with a low initial man-made capital stock and a high initial resource stock, the initial interest rate is higher than the discount rate. Under these circumstances it is efficient to invest in man-made capital so that consumption rises over time. However, as the stock of capital rises and the flow of resources from the non-renewable stock decreases, the interest

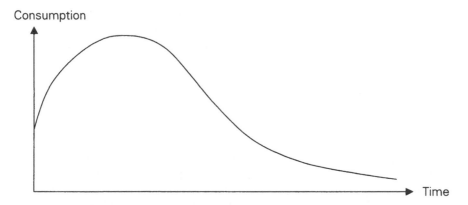

Figure 2.2 *The 'optimal' path of consumption in Dasgupta and Heal's model*

rate falls. At a certain point the interest rate falls below the discount rate. From this point onwards, consumption falls, as illustrated in Figure 2.2.

This model illustrates that, with a non-renewable resource stock and a man-made capital stock, it is possible for the optimal path to decline towards zero over time. This is a direct consequence of a positive (and constant) utility discount rate, which means that the cost of low levels of utility far into the future is outweighed by the benefit of higher levels of utility in the near future.

Sustainability and Natural Resource Use

The previous section has illustrated how both man-made and natural resources can be exploited over time so as to maximize the net present value of the path of utility, using a PV-maximizing criterion of optimization. In this section we will discuss how such a solution can be constrained by a criterion of sustainability. This means planning the use of natural resources and investment in man-made capital so that consumption does not decrease over time.

Sustainable extraction of a non-renewable resource is, for practical purposes, impossible. However, sustainable resource *use* can be defined as maintaining a capacity to provide the benefit in question – for example, by investing in a substitute for the non-renewable resource or, if the resource is exported, in an alternative revenue-generating asset. This corresponds to Solow's (1991) definition of sustainability as leaving to the future 'the option or the capacity to be as well off as we are'.

Sustainability and Non-renewable Resource Use

The best-known rule for sustainability in the presence of non-renewable resource extraction is the Hartwick rule (Hartwick, 1977). This sustainability rule builds on the contribution of Solow (1974) who was the first to prove the existence of a theoretical constant consumption path when a non-renewable resource is depleted. The context of Solow's model was a closed economy in which a non-renewable natural resource is combined in production with man-made capital. Such a constant consumption path is also the solution to a welfare function which seeks to maximize the welfare of the worst-off generation, among all present and future generations. Such a welfare objective is referred to as a 'maximin' function.

Hartwick (1977) used the same context to derive his original rule for investment for sustainable consumption, identifying the investment in man-made capital along the maximin path. Hartwick showed for this context that along the maximin path, investment in man-made capital is equal to the rents from the depletion of the non-renewable resource. Hence, Hartwick's rule has become known as the 'invest resource rents for sustainability' rule.

The Hartwick rule identifies constant consumption in the presence of non-renewable resource depletion. It ignores the possibility that the use of the resource has negative externalities and that well-being might be determined by factors other than consumption. Moreover, it was derived using the assumption that man-made capital can substitute indefinitely for the non-renewable resource. This can be made more realistic by assuming the existence of a backstop technology, so that for sustainability it is no longer necessary that the non-renewable resource stock last for infinite time. Under these assumptions the rule for sustainability still remains that resource rents are invested while the resource stock is depleted.

Sustainability and Renewable Resource Use

A discussion of sustainability is more intuitive in the context of renewable resources. Clearly, for a given stock of a renewable resource there is a level of exploitation, corresponding to the growth of the stock that is sustainable and does not involve depletion of the stock. Thus, the fundamental difference between renewable and non-renewable resources is that the potential exists for the resource stock itself to be managed sustainably. However, it is also true that if the objective of the owner of the resource is to maximise the monetary value of the stock, it may be that the optimal strategy is to reduce the size of the resource stock.

Figure 2.3 shows how, for a renewable resource stock, the efficient sustainable stock size is determined. The curved line represents the growth of the stock, given the stock size. As the stock increases, its growth increases, reaches a maximum and then decreases to zero once the stock reaches the 'carrying capacity' of its environment. The slope of the curve represents the marginal productivity of the stock. This is an important factor as it represents the return on holding the stock, rather than depleting it and investing the proceeds in an alternative asset. The interest rate represents the return on an alternative asset and therefore the opportunity cost of holding the resource stock. At all stock levels above S*, the return on the marginal unit of resource stock is less than the opportunity cost, and so it is efficient to deplete the stock up until this point.

Hartwick (1978) extended his sustainable non-renewable resource model to include renewable resources. He showed that for sustainable consumption, it is necessary for the rents from the depletion of the resource stock (which is extraction minus resource growth) to be invested in man-made capital. For maximum sustainable consumption to be achieved, resource extraction must follow the efficient path described above so that the stock is depleted until the growth rate is equal to the interest rate. If there is no positive stock level at which the rate of growth is equal to the interest rate, then the resource stock will be depleted

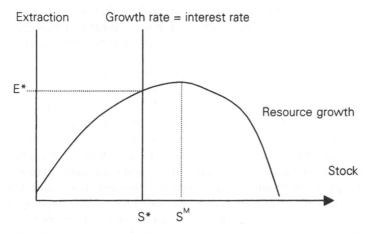

Figure 2.3 *Efficient resource depletion*

efficiently. Therefore sustainable consumption may well involve unsustainable resource extraction.

This discussion of sustainability has considered only sustainability of consumption, and the contribution of natural resources to production. Sustainability of consumption, however, is only one definition of economic sustainability. Arguably, economic sustainability also requires that the more fundamental services provided by the environment be sustained at or above some minimum levels over time. This can be described as environmental sustainability. The services that must be maintained if environmental sustainability is to be achieved are described in the next section.

The Environmental Dimension of Sustainable Development

The most conspicuous services that the natural environment provides are food and inputs to production, including energy, metals and timber. Energy and metals are non-renewable resources, while timber is a renewable resource. We have seen in the last section how these resources can be used sustainably in terms of maximizing the level of sustainable consumption that their owners can afford. In addition, the environment provides utility directly via, for instance, aesthetic and recreational values.

The natural environment also provides more fundamental services, without which human life on earth would not be possible. These are known as Global Life-Support services, since they provide the basic necessities to allow human life such as food and shelter, and the maintenance of suitable climatic and atmospheric conditions. There is often a trade-off between using natural ecosystems to provide inputs to production, even if this use is sustainable, and preserving them in their natural condition to maintain life-support services. The opportunity cost of preservation is the value that could be obtained from harvesting ecosystems and from converting the land to an alternative use.

Ecosystems, whether natural or managed, are required to capture the sun's energy and to produce food and raw materials. They regulate the hydrological cycle, which is a direct service to agricultural production. They create and maintain fertile soils, and they break down both natural and man-made waste into nutrients, maintaining soil productivity. Some of these ecosystem services are summarized in Figure 2.4.

A key life-support service provided by natural ecosystems is climate regulation, both on a local and a global scale. As noted, ecosystems are necessary at a local level for regulation of the hydrological cycle. On a global scale, they are necessary to maintain the composition of the world's atmosphere and therefore its system of climate regulation. The mitigation of global warming is the best-known example of this service. A major issue for global environmental sustainability is how far ecosystems can be depleted and pollutants allowed to accumulate in the atmosphere, without threatening global life-support services. One of the problems in analysing this issue is that scientific knowledge of ecosystem structure is far from complete. This results to a certain extent from the immense complexity and interdependence of species and systems. It is very difficult, therefore, to predict the effect that removing or reducing a population of one species will have on the rest of the species in a community or on the viability of the ecosystem.

Figure 2.4 highlights three important facts. First, the environment provides a wide variety of services that improve human welfare, both directly and indirectly. Second, the use of the environment for some services – for example waste absorption – reduces its capacity to provide other services – for example, amenity or resource inputs. Third, the services of the environment can be thought of as representing 'natural capital'. Stocks of natural resources, including the capacity to perform global life-support services, can be used unsustainably over time, reducing their capacity to deliver the same services in the future.

The implication of the second factor is that economic growth should be limited so that its negative environmental effects do not outweigh the positive contributions of economic growth to welfare. This is the point that Daly (1991) makes in complaining that while in microeconomics the optimal scale of an economic activity is clearly defined, with further growth uneconomic past the point at which the costs of growth outweigh the benefits, there is no corresponding recognition of an optimum scale of the macroeconomy. The question for environmental macroeconomics is, therefore, to what extent should economic growth be allowed to destroy environmental assets?

It has been suggested that this is a false dichotomy, since people become richer with economic growth and demand a cleaner environment. This hypothesis is encapsulated in the environmental Kuznets curve (EKC) debate (Grossman and Krueger 1991, World Bank 1992, Barbier, 1997), which suggests that the relationship between gross domestic product (GDP) and the quality of the environment is 'U-shaped' – ie, the quality of the environment deteriorates initially as GDP per capita increases and then improves after a certain critical value of per capita GDP has been reached. This critical value varies with the pollutant, and indeed for some pollutants such as volatile organic compounds (VOCs) there is no 'turning point'.

The EKC represents an empirically observed phenomenon – namely, the fact that some environmental problems have become less severe as income levels rise. Figure 2.5 illustrates the point, which has been used to argue that economic growth

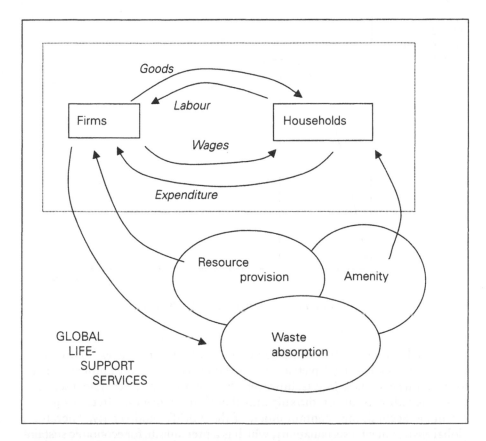

Figure 2.4 *The ecological services of the natural environment*

is, in fact, good for the environment, and that the argument that there are limits to the contribution that economic growth can make to human well-being is false. Although there is some merit in the EKC analysis it is not universally valid. The 'turning point' is not observed for all types of pollutants. For energy use, GHGs and waste, for instance, the relationship between incomes is increasing, rather than following the pattern depicted in Figure 2.5. The available empirical evidence is therefore insufficient to draw any general conclusions regarding the existence of an EKC effect.

A separate issue is the third fact mentioned above – namely, that the essential services provided by natural capital stocks may be run down unsustainably over time. If we accept that the market cannot reflect all the values of the services provided by the natural environment, this implies that environmental sustainability places a constraint on economic growth. The current section has underlined that natural ecosystems perform fundamental life-support functions without which human life would not be possible. However, because these services are not traded in markets, they are not reflected in the value of conserving natural resource stocks. Therefore, it is possible that when a natural resource stock is depleted efficiently in terms of maximizing its net present value, crucial life-support services could be lost.

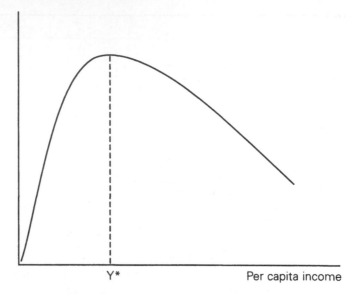

Figure 2.5 *An example of an environmental Kuznets curve*

An example of this conflict could be in the use of forest resources. Many of the world's forests are being depleted, since this is the economically optimal strategy for their owners. These forests, however, also perform global ecosystem services. First, they absorb carbon dioxide and therefore reduce the effects of global warming. Second, they contain much of the world's biodiversity. Therefore, global environmental sustainability, which is a prerequisite for economic sustainability, might require that limits be placed on the depletion of natural resource stocks, even if it appears that this means sacrificing some of the monetary value of these stocks.

The Social Dimension of Sustainable Development

In this section two lines of inquiry will be followed. The first extends the discussion of sustainability to include the concept of social capital. One reason for including social capital is the observation that economic and environmental factors fail to explain substantial differences in economic outcomes in countries with similar endowments of human, natural and man-made capital. Furthermore, similar sets of policies, such as structural adjustment, applied to countries with similar characteristics have resulted in diverging outcomes. This suggests that the three types of capital (human, man-made and natural) only partially determine the process of economic growth. By looking at the way in which the economic actors interact and organize themselves to generate growth and development, social capital is seen as providing a missing link (World Bank, 1997). Social capital is, however, very difficult to define and measure. Discussed below are the main aspects of social capital raised in the literature. Issues arising from the use of the concept of social capital to develop indicators are discussed in greater detail in Chapter 4. The general discussion about social capital leads to a second more specific inquiry about how the traditional literature on economic development has included issues of intragenerational equity and human development aspects.

The Concept of Social Capital

One of the reasons why the concept of social capital is intuitively appealing is that it encompasses a whole array of aspects concerning human interaction and the cultural and institutional foundations of society. This level of aggregation is, however, also the reason why there is a lack of consensus regarding the definition of social capital and why the concept is difficult to measure.

A frequently cited definition of social capital is that of Putnam whereby social capital has to do with those 'features of social organization, such as trust, norms, and networks that can improve the efficiency of society by facilitating coordinated actions' (Putnam et al, 1993, p167). With respect to the definition of development given below, social capital is concerned with the process by which inputs are transformed into well-being.

The divergences in the literature of social capital generally concern different views of which aspects of social capital should be emphasized. These differences are reflected in variations in the broadness of the definition of the concept. Furthermore, the validity of the word 'capital' to describe the concept has been questioned.[2]

The narrowest view of social capital regards it as a set of horizontal networks. The components of social networks and associated norms that influence productivity and well-being of a community are studied. The literature in this vein has focused on horizontal networks such as civic associations, rotating savings and credit associations, and credit cooperatives.[3]

A broader concept is to see social capital as consisting of both vertical and horizontal associations. It is then possible to study linkages transcending one community and thus consider the social capital aspect of integration with other institutions and/or levels of society. Studies of management of local common property resources and the resource allocating mechanisms involved are also found under this broader approach to social capital.[4] Apart from organizations and extended kinship, hierarchical relationships such as the Mafia and drug cartels can also be included in this broader view. Accordingly, the networks embodying social capital do not necessarily result in increased well-being.[5] This aspect is important since there has been a tendency to focus on the virtues of informal associations, sometimes neglecting the possibility that they may prevent more efficient resource allocating mechanisms from developing.

Common denominators of the mentioned approaches to social capital are that they all address cooperative engagements and the engagements take place within informal institutions at a level somewhere between the individual and the state. Furthermore, the issues of norms and trust are of central importance.

The broadest approach to social capital extends the concept to formalized institutions by adding the social and political environment and draws on the work of North (1990) and Olson (1982). The argument is that formalized institutional relationships and structures, such as government, the political regime, the rule of law, the court system, and civil and political liberties, significantly affect the rate and pattern of economic development (World Bank, 1997). It recognizes that the state and various interest groups are mutually dependent on each other: the capacity of various social groups to act in their own, and others', interest depends on the degree of support that they receive from the state and the private sector, while the state depends on social stability and popular support.

It can be argued that the different approaches to social capital are not so much alternatives as complementary dimensions of the same process (World Bank, 1997). The ultimate concern is to analyse the ways in which social relationships affect economic outcomes and economic efficiency, and the externalities created by these relations. Accordingly, the relevance of the various aspects of social capital for sustainable development is a matter of little disagreement. First, the formal and informal institutional setting has implications for the success or failure of the implementation of projects as well as policies. Second, the implementation of policies and projects may affect the institutional settings in the long run. Studies of the institutional setting can provide useful information for the design of both policies and projects.

The description of the three approaches to social capital illustrates the multiplicity of areas that need to be studied if social capital and the ways in which it affects economic outcomes are to be assessed, both theoretically and empirically. A few of these areas are mentioned here. First, the information-sharing role of institutions and more generally imperfect information should be studied. The implications for transaction costs can also be addressed. Second, cooperative action and the coordination of activities are central areas of study. The same is the case with a third possible area of study: informal insurance and its risk-sharing element. These areas are closely connected to the issue of trust. Furthermore, they all have implications, positive or negative, for the functioning and development of markets that are worth addressing.

The heterogeneity of the different forms of social capital makes it problematic to use an aggregate concept. If, however, we have to concentrate on specific elements of social capital depending on the context, it may be more rewarding to keep the focus on the institution, or resource-allocating mechanism, rather than on the type of social capital it is believed to embody.

The heterogeneity concerning both definitions and forms of social capital pose problems for measurement. To operationalize the concept it is necessary as a first step to be clear about what should be measured. According to the World Bank (1997), there is growing empirical evidence on the benefits of social capital, but few data on the cost side. This reflects the difficulty of determining the costs of creating an association or a certain level of trust. Similarly, studies of the potential negative implications of social capital expressed, for instance, by barriers to market penetration and the enlargement of networks are not readily available. However, an investment decision relating to social capital and the inclusion of social capital in project assessment requires a comparison of costs and benefits. The studies referred to above provide different solutions regarding measurement, but their contribution also lies in the provision of some preliminary indicators of social capital, discussed later in this chapter.

It is worth stressing another point made in Dasgupta and Serageldin (1999). Social capital cannot yet be viewed in line with human, man-made and natural capital. While markets are often imperfect and missing, market prices are usually present to offer a benchmark for estimating shadow prices. Social capital has its greatest impact on the economy in exactly those areas of transaction in which markets do not exist.[6] Until this problem is solved, aggregate social capital cannot be measured. This has serious implications for the development of indicators that capture the social capital aspect of sustainable development (it is possible to argue that the measurement of natural capital faces much the same problems, but the difficulties are less severe).

Economic Development, Equity and Human Development

Although a generally accepted definition of development does not exist, there is broad consensus that it implies increasing social welfare or well-being.[7] At the individual as well as aggregated level, increased welfare can be seen as reflecting both access to the relevant inputs and the ability to transform these inputs into well-being. Among the multiplicity of factors determining this, the issues of economic growth (which were addressed earlier in this chapter), equity, basic human needs and freedom are predominant in the literature addressing the human aspects of development. All these issues may be seen as different dimensions of, or components of, inter- and intragenerational equity. In the remainder of this section intragenerational equity is discussed further from three perspectives:[8] (a) the economic perspective, (b) the environmental perspective, and (c) the capability perspective. This is followed by a remark about human development indicators.

The Economic Perspective

Development is broadly defined as increasing social welfare. In development economics, social welfare is generally assumed to be positively affected by the income level and negatively affected by inequality in the distribution of income.[9] Accordingly, the economic approach to assessing intragenerational equity focuses on the level of income and the distribution of income. While it is recognized that poverty and income distribution are incomplete indicators of the human aspects of development, the observed correlation between income and other indicators of human development (discussed below) suggests that income is an important indicator of many aspects of development and well-being (see, for example, Fields, 1980).

With respect to sustainable development, the income distribution between and within countries and regions is one area of interest. Measures of income inequality fall into four main types: absolute and relative income measures, and absolute and relative poverty measures. Income inequality measures are not objective and different measures can give diverging answers regarding the trends in inequality. The choice of income inequality measure(s) and the question of weighting should therefore depend on which aspects of inequality are considered important in a given context. Typical measures will include an absolute income measure such as GDP and an assessment of relative income inequality – for instance, by ranking household per capita incomes according to size. Furthermore, measuring the number of persons who fall below a defined poverty line over a period of time is one way of assessing the development in absolute poverty.[10]

One advantage of focusing on income is the fact that it provides a highly operational measure for which data are relatively easily available. However, the focus also has a number of shortcomings. If well-being is seen as reflecting access to the relevant inputs and the ability to transform these inputs into well-being, as argued above, the income approach only reflects access to one input, ignoring other inputs as well as the transformation process by equating income and well-being. In part this can be corrected; inputs other than income may be included and represented by a monetary value (household surveys generally include the assessment of non-marketed activities). In resource-based subsistence economies such activities are especially important to include, as they can be a substantial fraction of the household budget. In relation to sustainable development, using income as an indicator may, however, hide information about resource use and the access and availability of resource that can be a central area of study. Some of these issues are discussed further below.

The Environmental Perspective

It was made clear earlier in this chapter that we all depend on the natural resource base for sustaining our lives. The direct dependence on the natural resource base is, however, even more pronounced for many developing countries and especially the poorer parts of the population in these countries. This can be illustrated by looking at the contribution of the natural resource base to GDP, which is generally larger in developing countries (World Bank, 1997). The lack of substitution possibilities (such as the reliance on fuelwood) and restraints on technical options (such as wells or methods of water purification) for mitigating the direct dependence on the natural resource base in many cases will be more binding in developing countries. In line with this, Pearce, Barbier, and Markandya (1990) note that non-declining natural capital stocks can generally be assumed to benefit the poor in developing countries.

The global distribution, or prevalence, of natural resources is far from equal. Apart from reflecting climatic and geographical differences, it also reflects that some regions have depleted the stock of natural capital and/or substituted it for other types of capital. Likewise, the nature and extent of polluting activities differ across regions, reflecting, among other things, the size and structure of the economy. These differences are part of the rationale for international environmental agreements, but also point to the problem of fairness in burden sharing when designing these agreements.

From a development and equity perspective the overall availability of natural resources is only a partial indicator. It has to be supplemented with information about the relative availability of and access to natural resources.[11] One commonly used indicator that reflects both the access and availability aspect is time spent collecting, for instance, fuelwood and water. Other indicators may be in the form of actual consumption by different groups (for example, by rural or urban location or by income) of various environmental goods (including clean air).

The Capability Perspective

According to Sen (1992), an assessment of inequality should acknowledge two fundamental types of differences: the various variables that can be used for assessing inequality and the basic heterogeneity of people. Using income as the basis for assessing inequality ignores these differences. The first difference is ignored for obvious reasons, since only one variable is considered. The second difference refers to the fact that we vary in our ability to transform the inputs (or means) into what we want (our ends), which the use of, for example, an aggregate social welfare function ignores. Furthermore, one unit of income may not be homogeneous because of regional differences, just as the distribution of income does not reflect the distribution of consumption of social services or distortions in the distribution of goods and services.

In the capability approach, inequality reflects that there are factors resulting in differences in what people can be and do. Consequently, inequality is closely related to freedom.[12] Central in the capability approach is the notion of functionings, defined as 'beings and doings' (Sen 1992, p39) and covering a wide spectrum. Examples of functionings include being healthy, getting adequate nutrition, taking part in the activities of society and having self-respect. The difference between capability and functioning is that while a functioning can be regarded as an element, capability refers to a set of such elements, reflecting what a person can be or do and thereby also reflecting the limitations of the possibilities facing the person. In this way, the two concepts can be said to address different dimensions of the inequality aspect: the functionings achieved are related to the achievement of well-being and the capability to function is related to the freedom to achieve well-being.

Apart from acknowledging the differences mentioned above, the capability approach has the potential for making the assessment of inequality more objective, given the set of capability indicators selected to be relevant for a given policy evaluation, than it is in the utilitarian-based income approach. The utilitarian income approach involves the comparison of welfare or utility, which Sen describes as a mental characteristic. While the capability approach involves a certain amount of subjectivity in the selection of the functionings and capabilities, the assessment can be made objectively by comparing the functionings achieved by different persons, or by comparing the differences in capability sets. The approach is not readily operational since it involves large amounts of data and offers no aggregated indicators.

Indicators of Human Development

The reductionism of using income as an approximation of well-being is well recognized. Much in line with the capability approach, some have suggested the satisfaction of basic human needs such as minimal levels of nutrition, health, shelter and opportunities for individual freedom, as an alternative. In practice, rather than dismissing the economic approach to inequality, it is supplemented with other indicators of development such as the indicators of basic human needs already mentioned, life expectancy at birth, adult literacy, fertilizer use and energy consumption. The indicators may be analysed separately or integrated into a social welfare function. It is problematic, however, to aggregate the indicators to reflect overall well-being.

A wide range of social indicators can be found in the Human Development Reports and Human Development Indicators published by the United Nations

Development Programme (UNDP, 1999). They also present a Human Development Index, which is an example of an aggregated indicator as mentioned above. The World Bank also publishes social indicators.[13]

Measures of Sustainability

It is clear from the discussion so far that there are various types of capital stock that contribute to human well-being. These include man-made capital, such as factories and machinery. Human capital is also a productive stock that can be invested in and allowed to deteriorate. An intangible type of capital is social capital, discussed above. Finally, there is natural capital which performs many different functions. The total capital stocks can be seen as representing the wealth of a given economy. The reason for focusing on capital or wealth in relation to sustainable development is that it is perceived as the basis for meeting the needs of present and future generations. Put differently, the total capital stock is the basis for generating income flows and other flows related to our well-being.

Weak and Strong Sustainability Concepts

The fact that there are different types of capital stock that contribute to well-being has led to a distinction between weak sustainability and strong sustainability, as discussed by Rennings and Wiggering (1997). Weak sustainability can be defined as the maintenance of the value of the aggregated stock of capital. This implies two things. First, that different capital stocks can be expressed in common terms – namely, in terms of monetary value. Second, that different types of capital stock can substitute for each other in a sustainable solution. A definition of weak sustainability might be:

$$\dot{K} + \dot{H} + \dot{SC} + \dot{N} \geq 0$$

where K is man-made capital, H is human capital, SC is social capital and N is natural capital. As long as these capital stocks can be expressed in money terms, then under a policy of weak sustainable development, depletion of the stock of natural capital may be compensated for by investment of the same or greater value in, for instance, man-made capital.

Strong sustainability, on the other hand, requires that each type of capital stock be maintained in its own right, at least above some minimum level. Therefore, a strong sustainability constraint might be expressed as follows:

$$\dot{K} \geq 0, \, \dot{H} \geq 0,$$

$$\dot{SC} \geq 0, \, \dot{N} \geq 0$$

Further disaggregation is required where different aspects of the natural capital stock do not substitute for one another. For example, increased fish stocks may not compensate for deforestation. Thus, in the case of natural capital the minimum level of different types of natural capital stock might be determined by safe minimum standards and the Precautionary Principle. Under a policy of strong sustainability, the depletion of the world's forests might be limited so that the remaining stock is sufficient to maintain biodiversity at safe levels, to sustain timber supplies at adequate levels and to absorb carbon dioxide emissions. In the rest of this section we look at the measurement of strong and weak sustainability at the macrolevel. Much of the literature has been devoted to such measures and they are most closely tied to the conceptual basis outlined above. Microlevel strong and weak sustainability indicators applied to GHG emission-reduction projects are discussed on pp39–44.

Daly's Rules for Strong Sustainability

Herman Daly's (1990) criteria for sustainability provide an example of a suggested programme for strong sustainable development. Daly's criteria are as follows:

1   Renewable resources must be harvested at or below the growth rate for some predetermined level of resource stock.
2   As non-renewable resources are depleted, renewable substitutes must be developed to maintain the flow of services over time.
3   Pollution emissions should be limited to the assimilative capacity of the environment.

In another paper Daly (1995) answers critics of sustainability, such as Beckerman (1994) by noting that it does not require, as Beckerman maintains, every species to be preserved and that no non-renewable resource can ever be extracted. Rather it involves acknowledging, in contrast to the weak sustainability paradigm, that natural resources are basically complements to man-made capital in production and that the decreased availability of natural resources can be compensated only to a limited extent by increased man-made capital. Therefore Daly does agree with Beckerman that the concept of weak sustainability is illogical, since it assumes that man-made capital can substitute for natural capital, an assumption which Daly says is not borne out empirically.

Indicator of Weak Sustainability: Genuine Savings

An example of an indicator of weak sustainability is provided by Atkinson and Pearce's (1993) measure of 'Genuine Savings'. This indicator takes a country's savings and deducts from it the value of the depreciation on its man-made capital and the value of the depreciation on its natural capital. This allows a judgement to be made regarding a country's sustainability, since if the genuine savings measure is positive, then the country is weakly sustainable. The genuine savings measure is the following:

$$Z = \frac{S}{Y} - \frac{\delta M}{Y} - \frac{\delta N}{Y}$$

where Z is genuine savings, Y is income, S is savings, δ is the depreciation on M, the country's stock of man-made capital, and on N, the country's stock of natural resources. Table 2.1 provides a summary of the empirical study undertaken by Atkinson and Pearce (World Bank, 1997).

These estimates have been criticized for the same reason that Daly criticized the concept of weak sustainable development – ie, it is illogical to suppose that maintaining the overall level of capital stock, with increases in man-made capital substituting for decreases in natural capital, can constitute a policy of sustainable development. Moreover, countries such as Japan, which deplete very few natural resources, can count most of their investment, net of man-made capital depreciation, as genuine savings. However, their production is dependent on natural resource imports, which may well derive from unsustainable extraction. Therefore, the high sustainability rating for such countries may be misleading. Atkinson and Pearce's position is that this measure does not necessarily mean that a country is sustainable, but rather that if a country does not pass even such a weak test of sustainability as this, it is a clear sign of *un*sustainability.

Genuine savings can be thought of as one aspect of the adjustments to a country's Net National Income (its output plus its *net* investment) that should be made to account for the services provided by the natural environment described on pp21–24, and any depletion of the capital stock that these services represent.

Rule for Strong Sustainability: The Shadow Project Constraint

Another rule for strong sustainability, suggested by Pearce et al (1990), is the shadow project constraint. The idea, based on strong sustainability, can be interpreted as requiring the stock of natural capital, including ecosystems, not to decrease over time. This is quite a stringent criterion since it implies, for example, that no new land may be developed. It could mean that a development project with large economic benefits might have to be forgone because it involves some environmental damage. The shadow project constraint presents a way to avoid this problem. The standard cost-benefit criterion, including environmental costs and benefits is:

$$\sum_{t=0}^{T} B_t \theta_t - \sum_{t=0}^{T} C_t \theta_t - \sum_{t=0}^{T} E_t \theta_t > 0$$

Where T is the time period over which the costs and benefits of a project are analysed, θ is the discount factor in each period, B is monetary benefits of the project, C is its monetary costs and E is its environmental costs. Therefore the standard cost-benefit criterion is that the net present value of the benefits of a project must outweigh the net present value of its costs, including the environmental costs. The shadow-project constraint allows for any ecosystem destruction

Table 2.1 *Measures of genuine savings for selected countries (% of annual GDP)*

	$\dfrac{S}{Y}$	$-\dfrac{\delta M}{Y}$	$-\dfrac{\delta N}{Y}$	z
Sustainable economies				
Brazil	20	7	10	+3
Costa Rica	26	3	8	+15
Czechoslovakia	30	10	7	+13
Finland	28	15	2	+11
Germany	26	12	4	+10
Hungary	26	10	5	+11
Japan	33	14	2	+17
Netherlands	25	10	1	+14
Poland	30	11	3	+10
USA	18	12	3	+3
Zimbabwe	24	10	5	+9
Marginally sustainable				
Mexico	24	12	12	0
Philippines	15	11	4	0
United Kingdom	18	12	6	0
Unsustainable				
Burkina Faso	2	1	10	−9
Indonesia	20	5	17	−2
Madagascar	8	1	16	−9
Nigeria	15	3	17	−5
Papua New Guinea	15	9	7	−1

Source: World Bank, 1997

to be compensated by a shadow project which increases environmental quality elsewhere. Given this, as well as the standard cost-benefit criterion, the project must satisfy:

$$\sum_{t=0}^{T}\sum_{i=1}^{n} E_{it}\theta_{t} \leq \sum_{t=0}^{T}\sum_{j=1}^{m} a_{jt}\theta_{t}$$

where there are i sites at which environmental damage is caused and a denotes the j shadow projects that create environmental benefits. Therefore this condition states that the net present value of environmental damage over the relevant time period must be negative. This is a weak version of the shadow project constraint. A strong version would be that the environmental damage in each time period must be negative.

Rule for Strong Sustainability: Safe Minimum Standards

Another rule, associated with land use and non-declining natural capital stock, is the safe minimum standards (SMS) approach. Ciriacy-Wantrup (1952) and Bishop (1978) developed this approach. It stems from a concern that the type of calculation carried out under cost-benefit analysis cannot be used to plan for sustainability, because the valuation of damage to ecosystems cannot reflect sustainability principles. In the absence of a reliable calculation, it is suggested that ecosystem damage be limited so that the remaining stocks are above safe minimum levels, usually calculated as the minimum levels required for the ecosystem to remain viable.

The SMS rule, therefore, is 'prevent reductions in the natural capital stock below the safe minimum standard identified for each component of this stock unless the social opportunity costs of doing so are "unacceptably" large'. (Hanley et al, 1997). This implies, for example, that pollution emissions and biodiversity loss should be kept below identified safe levels. The indicator of sustainability implied by this criterion is then whether or not the SMS is breached for any class of resource.

MAKING SUSTAINABLE DEVELOPMENT OPERATIONAL

Sustainable Development Criteria in the Context of GHG Emission-reduction Projects

Studies of GHG emission-reduction projects are often structured to assess how specific projects meet a given decision criteria, which is often based on the costs and benefits of implementing the projects. Since climate change is a global environmental control problem,[14] a given GHG emission-reduction policy seen from a global perspective is most cost-effective if marginal emission reduction costs are equalized across emission sources. Seen from a national policy perspective, it can be argued in the same way that it is economically efficient to implement a given GHG emission-reduction target with the policies and options that have the lowest marginal reduction costs irrespective of the location of the sources. Given these general arguments, a common measure for the effectiveness of GHG emission-reduction policies is the cost of implementing a given project per unit of GHG emission reduction.

However, most GHG emission-reduction projects have a number of indirect impacts on the economic, social and environmental development, which reflect important aspects of sustainable development. These multiple impacts raise the issue of how such projects should be evaluated.

The approach for an integrated assessment of GHG emission-reduction objectives and sustainable development policy objectives should reflect how the policy objectives are structured and prioritized. Two different cases can be distinguished, namely:

1   Where GHG emission reduction is the main policy objective and sustainable development impacts are considered as an indirect impact of the policy.

2 Where GHG emission reduction and sustainable development policies are considered as joint policy objectives.

In case 1 it is best to start with an assessment of the costs of a range of GHG emission-reduction projects and then, as a subsequent step, to assess indirect impacts on sustainable development. The 'decision rule' for these projects is to select the projects that have the lowest cost per unit of GHG emission reductions, subject, however, to the requirement that the projects do not have 'unwanted' indirect impacts on sustainable development. Various rules for 'judging' these indirect impacts on sustainable development can be suggested as described on pp30–34 on measures of sustainability.

For policies of the kind described under case 2, a different decision rule for project selection is required. The direct costs of implementing a given GHG emission-reduction project in this case should be integrated in the assessment of sustainable development impacts, and the project selection rule in this case directly should reflect the total sustainable development impacts of the projects.

In this way, the first project-screening criteria will be the rules for weak and strong sustainability like the ones that are outlined p30–34. Given that a number of projects meet specific sustainability rules, the actual prioritizing of projects again can be done on the basis of different decision criteria which, for example, can be that projects with the highest total sustainable development impact are selected.

Application of Sustainability Rules to GHG Emission-reduction Projects

The aggregate measures of strong and weak sustainability adapted to global ecosystems which were outlined on pp30–34 are likely to be of limited use in relation to the evaluation of GHG emission-reduction projects, most of which will be at the microlevel. Nevertheless, we believe it is useful to track strong and weak sustainability for the economy and this can be done at the microlevel.

Specifically for strong sustainability we can adapt measures based on:

1 Daly's rules for strong sustainability.
2 The shadow project concept.
3 The safe minimum standard.

We consider each of these below.

Daly's Rules for Strong Sustainability

This rule can be adapted for use at the project level as answers to the following questions:

1 Does the project's use of renewable resources exceed the rate of natural growth?
2 Do the pollution emissions from the project take total emissions over the critical load?

The Shadow Project Concept

This could be used in a GHG emission-reduction project to demand that any important loss of natural capital be compensated by investment in a shadow project that creates a natural resource of similar value.

The Safe Minimum Standard

This defined natural physical limits to the use of assimilative capacity and ecosystems. The question that follows from it is partly (b) under Daly's rules for strong sustainability. In addition, the criterion requires that depletion of bio-diversity and other natural assets should not exceed the SMS.

We have included indicators based on the above on pp39–44.

For weak sustainability, the contribution of a project to genuine savings is given by:

> Gross domestic investment in project
> minus investment displaced elsewhere
> minus additional depreciation of natural capital
> minus additional depreciation of man-made capital.

Note that this measure will be dynamic – ie, it will cover more than one year. We include indicators based on this concept on pp39–44.

Approaches for Designing and Selecting Specific Policy Objectives for GHG Emission-reduction Programmes

The decision-making framework that is used to select sustainable development impact indicators and value these should reflect the specific priorities and per-spectives of the parties involved in the policy implementation. In this way, the decision-making framework should facilitate a social decision-making process where the many different aspects of sustainable development impacts related to the GHG emission-reduction projects are considered in a consistent and trans-parent approach.

As addressed earlier on pp16–30, the scope of sustainable development impacts that are relevant to consider goes beyond what can be valued on the basis of individual preferences or in a social welfare function. It is therefore recom-mended that the actual selection and evaluation of sustainable development impacts in relation to specific GHG emission-reduction projects are related to a broader social decision-making process that, for example, can include expert judgements or decision-making processes that involve various stakeholders (Cantor and Yohe, 2000).

One approach to establishing a social decision-making framework is to use social preferences as they have been revealed by past political decisions. Such political decisions can be reflected, for example, in national development pro-grammes, or in specific sectoral or social programmes. The programmes then can be used as a background for developing a reference scenario for the policy evaluation and for selecting sustainable development impact indicators that are considered to be important policy objectives in the programmes. Another approach is to structure decision-making processes where experts or stakeholders are invited to discuss and select policy objectives and prioritize them.

The remaining part of the chapter covers issues concerning the operationality of sustainable development in relation to GHG emission-reduction projects. Assuming that a definition of sustainable development is established, the next step is to make the concept operational by developing criteria for policy assessment. As mentioned earlier, we take the position that there are strong arguments for basing a sustainable development criterion on host country priorities that reflect national development perspectives.

A process for policy assessment involving the following three steps is likely to be effective. The national development plan serves as the starting point. This plan identifies the main development goals, their relative importance and places them in a timeframe. Furthermore, it gives an indication of which areas within the three dimensions of sustainable development – ie, the economic, the environmental and the social dimensions – are considered most important.

The second step is to investigate the possible GHG emission-reduction projects. This step includes identifying the types of projects, assessing the main areas of impact and the expected significance of these impacts, and analysing possible spillovers (or externalities) of the project.

The third and final step of the process is to compare the impacts of the possible projects and assess their consistency with national development priorities.

The establishment of an agreed conceptual and operational framework is a requirement for steps two and three. The analytical approach should be internally consistent, and assessment methods and indicators should support each other. The choice of indicators in the different dimensions of sustainability should depend therefore on the choice of assessment method. In the same way limited data availability for indicators may prevent the use of certain assessment methods. Described below are various assessment methods that can be used in relation to GHG emission reduction projects, followed by a range of indicators that can be of use in such assessment. Before discussing the available asessment methods, we discuss some of the conclusions that can be reached regarding operational rules for the assessment of projects as described in steps one to three above.

Operational Rules: The Economic Dimension

The issues addressed on pp17–21 regarding the economic dimension of sustainable development reflect matters of importance when building a national development plan. Intergenerational equity aspects are influenced by the choice of a development path, just as decisions regarding natural resource use, substitution and depletion can be seen as reflecting overall development policy perspectives. At the project level conclusions can be drawn concerning the economic soundness and efficiency of the project, and on the implications of key economic policy objectives such as employment generation, balance of trade, economic growth,and specific sector objectives.

Operational Rules: The Environmental Dimension

Operational sustainability criteria applied in relation to projects are complicated to derive owing to the fact that sustainability, by definition, applies to a whole system and it can be difficult to assess how a specific project contributes to this. There are different approaches to establishing a 'link' between sustainability criteria at the systems level and project assessment criteria. These include definition of resource stock constraints and shadow projects, as well as the inclusion of a transfer component in a project to compensate for long-term environmental impacts that have an intergenerational equity dimension, such as long-term climate change damages.

Operational Rules: The Social Dimension

One approach to addressing income distributional consequences is to map all policy impacts in relation to income groups and hence to estimate the distributional incidence of control costs and environmental benefits. The incidence of control costs can be assessed, for example, as a study of tax expenditures of household income segments if environmental taxes are applied based on detailed household expenditure data. The environmental benefit study can assess the actual individuals who are exposed to the particular pollution and the benefits for these groups of reducing the exposure based on pollution damage estimates and consumers' preferences for the actual environmental good. Likewise, the effects on employment, health, education, access to natural resources, and so on can be assessed at the project level.

The overall distributional effects of any one project can be expected in most cases to be limited due to the small scale of most activities. However, some projects can involve significant gains and losses for specific stakeholders, which should be taken into account in an assessment of the social impacts of project implementation. Examples of projects with such significant impacts are projects that substitute fuels that are domestically supplied, such as coal-based activities or woodfuel use. A GHG emission-reduction project that replaces such activities as a side-impact imply decreasing employment in the fuel-supply sector as well as decreased revenues to the fuel suppliers.

It is more difficult to conduct intergenerational studies of distributional consequences of environmental benefits. Such studies will need to be based on forecasts of environmental loads over time, the relative prices of environmental goods and other commodities, and assumptions about the preferences of future generations.

The project contribution to and implications for social capital can be based on considerations such as the extent to which the project facilitates establishing networks, the exchange of information and knowledge, the linkages between local networks and other levels of society, participatory decision-making, the development of markets and social mobility. Another indicator of a positive social impact of a project implementation is that the project includes a social learning component. The inclusion of social leaning can imply, for example, that the project leads to the establishment of capacity (information, skills, horizontal networks in the management) for managing similar projects, that will decrease their implementation cost over time.

The social implication of a project can also be assessed, as argued on p29, in relation to capabilities. Indicators for measuring capabilities for individuals include energy access, education, time, nutrition and health.

Indicators for Measuring Sustainability in the Context of Projects

In this section we develop specific indicators that are likely to be of use in GHG emission-reduction projects. The key questions asked about the indicators are:

1   Are the data available for measuring the indicator and if so at what cost? A scale of 0–3 is used for the answer. A '3' means that data are very easy to collect and always available. A '0' means that data will have to be collected specifically and at high cost. Moreover, it may not be feasible to provide the indicator in some countries.
2   How good is the theoretical basis for the indicators? The same scale is used, with a 0 indicating that there is no theoretical basis for the indicator – it is totally ad hoc. A 3 signifies that a full theoretical discussion is available in literature.
3   Does the indicator have a time dimension? This is scaled as Low (L), which means that a single value will probably suffice, to Medium (M), which means that future values may vary from the present ones, to High (H), implying that it is critical to track the indicator for a number of years of the project.
4   Is the indicator relevant to the projects? Here the same 0–3 scale is used, with 0 indicating that the indicator is not relevant, to a 3 indicating that it is highly relevant.

Finally, we indicate the sectors where the indicators are most likely to be relevant. Table 2.2 deals with environmental sustainability indicators and Table 2.3 with economic and social indicators. This selection is based on an extensive review of the indicator literature, with the key references included in the bibliography.

These indicators cover a wide range of impacts and it is not expected that all projects will use all of them. One of the areas of further work is to provide more guidance on where particular indicators should be used. The tables provide, however, a useful 'shopping list' of indicators, based on the most recent literature on the subject. Finally, Boxes 2.1 and 2.2 report indicators of social acceptability and social capital that have been applied to project assessment.

As a general point, specific indicators will apply in specific cases and we cannot offer generic advice for all projects. Nevertheless, some general conclusions can be drawn from these tables:

1   The environmental indicators of greatest value are
    • Changes in water and air pollutants accompanying the GHG mitigation project
    • Changes in the use of raw materials and minerals accompanying the project
    • Changes in sustainable or unsustainable forestry practices induced by the projects
    • An important indirect indicator is the impact of the project or policy on the use of renewable energy in the country
2   The time dimension is important in most of the key indicators.
3   On the socioeconomic indicators the most important direct impacts are those on employment and food supply (in sequestration projects). The most relevant indirect impacts are on productivity and regional value added. A useful

Table 2.2 *Environmental indicators of sustainability*

Indicator	Data availability/ sources	Theoretical basis	Time dimension	Relevance to GHG projects	Sectors where relevant
	A	B	C	D	
Land converted out of natural resource use[a]	1	2	M	2	Forestry, biomass, hydro, transport
Increase in ratio of renewable to non-renewable energy in region	2	2	H	3	All renewable energy projects
Contribution to unit of renewable energy for other projects in the country	3	2	H	3	Most renewable costs energy projects
Contribution to unit costs of non-renewable energy for other projects in the country	3	2	H	3	Several non-renewable energy projects
Impact of project on water quality in the region	2	2	M	2	Many energy projects where water is used and released
Impact on hydrological functions for water	2	2	M	1–2	Afforestation projects and hydro

Impact on soil erosion in the area of the project	2	3	H	1–2	Hydro, afforestation
Change in the land affected by desertification	2	3	H	1–2	Hydro, afforestation
Change in the emissions of pollutants (sulphur and nitrogen oxides, ozone-depleting substances)	3	3	L	3	Energy efficiency projects, transport
Change in the rate of waste recycling	3	3	L	3	Biomass projects, waste as energy
Change in the rate of water consumption and withdrawal	2	2	M	2	Energy efficiency projects and land conversion
Impact on natural marine ecosystems (fish, etc)	1–2	3	H	2	Hydro energy projects
Impact on wildlife habitat and biodiversity	1–2	3	H	3	Afforestation, hydro, transport
Timber harvest to growth ratio	2	3	H	3	Afforestation projects

Note: a We may wish to report separately loss of agricultural from other types of land (eg forestry)

Table 2.3 Social and economic indicators of sustainability

Indicator	Data availability/ sources A	Theoretical basis B	Time dimension C	Relevance to GHG projects D	Sectors where relevant
Change in the mortality rates	0	1–2	H	1	Energy efficiency projects, transport
Changes in the income of the population, income distribution	2	3	L	3	Most projects
Change in the rate of unemployment	3	3	M	3	All sectors
Change in the percentage of women employed	0–1	2	H	2	All sectors
Change in number of years of training	1	3	M	2	All sectors
Impact on public participation[a]	1	2	H	2	Transport
Contribution of the project to genuine savings in the economy[b]	1	3	H	2	All sectors with investment and natural capital
Changes in productivity outside the project as a result of the project	0–1	2	M–H	2–3	All sectors
Indirect value added in the region as a result of the project	3	3	M	3	All sectors

					Relevant projects
Change in the access to public transportation	2	3	M	2	Transport
Sense of individual empowerment[a]	0	1	M-H	1-2	All projects with additional employment
Impact of the project on regional/national budgets	2	1	M	1-2	Projects with significant employment effects
Impact on food supply in the region (eg sink project that reduces food security)[c]	2	3	M	3	Land conversion projects (afforestation, agriculture), transport
Change in the access to health care[d]	1–2	3	M	2	Transport
Change in the access to education[d]	1–2	3	M	2	Transport
Changes in noise from road traffic	2	3	M	2	Transport
Changes in the rate of transport fatalities	1	2	H	1–2	Transport
Change in the rate of production-related injuries	1	2	H	1–2	Industry-related projects

Notes:
a These indicators are qualitative only
b Construction of the measure is discussed in the text
c Food supply may be reduced with afforestation programme, but improved with better public transport
d These indicators pick up the indirect impacts of improved transport

innovative indicator is the contribution of the project to genuine savings, although this has yet to be implemented.
4   The indicators of social acceptability and social capital are not *indicators* in the usual sense of the term and they are difficult to link to individual projects. Their main usefulness is as a guide to appropriate prior 'operating conditions' for the projects. Ideally, we should ensure that these conditions (for example, a management plan, high social mobility, good horizontal networks and so on) are present. The question then arises, what if they are not? Projects undertaken in the absence of what we will call 'good reference operating conditions' will need further support if they are to be successful. This issue is raised again later in the book.

CONCLUSIONS

This chapter has provided an overview of the problem of sustainability and some of the ways in which sustainability is analysed in the literature. It has also discussed the extent to which sustainability can be measured and the measures used in assessing GHG emission-reduction projects.

We have seen that the economic, environmental and social factors that contribute towards human well-being can be thought of as being 'capital' stocks. These stocks can be used sustainably, so that the implicit or explicit interest on the stocks can be consumed while the overall stocks are left intact. Alternatively, they can be used in an unsustainable way, which means that they are allowed to deteriorate or be consumed over time. There are two reasons why the use of these capital stocks might not be sustainable. The first is where people prefer to increase their own current consumption at the expense of future consumption. This tendency can be expressed as a high rate of discount of future utility. The second is where the services provided by these capital stocks are not fully accounted for in decision-making, so that decisions are taken that result in the deterioration of essential capital stocks. We have discussed examples of this phenomenon, which is particularly likely to apply to environmental and social capital.

The two main frameworks within which economic sustainability has been analysed can be described as weak sustainability and strong sustainability. The difference between these two paradigms is that weak sustainability assumes that different forms of capital can be substituted for one another along a sustainable path. Strong sustainability, on the other hand, assumes that substitution between, for example, man-made capital and natural capital is very limited. We have discussed some of the rules and indicators that have been developed in the literature for both weak and strong sustainability.

Applying these concepts at a local level is possible but requires the measures to be made operational. We have indicated how the concepts of strong and weak sustainability can indeed be made operational in the general context of a project, and we have used that discussion to reference a wide range of indicators, most of which are in the literature on sustainability at the local level, but some of which are not. The tables for operational indicators need some further development, showing how they can be constructed and what issues arise in applying them. This is done in other chapters of this book.

BOX 2.1 INDICATORS OF SOCIAL ACCEPTABILITY

Ecological and environmental soundness
- management plan, environmental management and monitoring
- municipal and village resolution endorsing the project
- endorsement letters from local NGOs and People's Organizations

Effective implementation of the public participation process
- process documentation reports signed by stakeholders
- scoping report signed by all key parties and stakeholders' representatives
- detailed description of the Environmental Impact Assessment process with concurrence of all stakeholders who participated
- signed Memorandum of Understanding for the establishment of a multi-partite monitoring team
- report of hearing officer during public hearing

Resolution of conflicts
- negotiated agreements on conflicts should be included in Memorandum of Understanding between the proponent, the Department of Energy and Natural Resources, local government units and stakeholders
- a resettlement and relocation plan
- social development programme

Social and intergenerational equity and poverty alleviation
- environmental and monitoring plan which includes a social development programme, compensation and resettlement plans
- endorsement letters from local NGOs and People's Organizations
- municipal and village resolution endorsing the project

Note: Based on criteria suggested for the Philippines by DENR, 1997

Box 2.2 Indicators of social capital

Horizontal networks
- number and type of associations/local institutions
- extent of membership
- extent of participatory decision-making
- extent of homogeneity within the network with respect to, for example, kinship, occupation and income level
- reliance of networks of support
- percentage of household income from remittances
- extent of linkages with other institutions/associations

Civil/political society
- index of political discrimination
- index of extent of democracy
- index of corruption
- strength of democratic institutions
- constitutional government changes
- degree of decentralisation of government
- coups
- percentage of people facing economic discrimination

Social integration
- indicator of social mobility
- ethnolinguistic fragmentation
- riots/protest demonstrations
- crime rates
- strikes
- divorce and suicide rates

Legal and governance aspects
- quality of bureaucracy
- expropriation and/or nationalization risk
- repudiation of contracts by government
- contract enforceability
- independence of court system

Note: Based on World Bank, 1997

Notes

1 This is more popularly known as the Brundtland Report, after the chair of the commission.
2 Dasgupta and Serageldin (1999) note that with respect to both heterogeneity and intangibility, social capital resembles knowledge and skills, which are regarded as forms of capital by economists.
3 The literature includes Narayan and Pritchett (1997), Putnam, Leonardi and Nanetti (1993) and Banerjee, Besley and Guinnane (1994).
4 See, for example, Bromley and Feeny (1992) and Baland and Platteau (1996).
5 See, for example, Coleman (1990), Olson (1982), Putnam et al (1993) and Dasgupta (1993).

6 To a certain degree, the measurement of natural capital poses the same problems. Contingent valuation methods seek to include option and existence value for which markets are not available to offer benchmarks.

7 Well-being – for an elaboration of the concept, see, for instance, Dasgupta (1993) – is generally a less controversial and more appropriate term to use than, for instance, welfare. Alternatively, fulfilment of needs or increasing living standards are terms that could be used here.

8 Intergenerational equity was addressed on pp17–21 and will not be mentioned in this part of the chapter.

9 Although this is specific to the welfare function applied, see below.

10 For thorough analysis of income inequality issues, see, for instance, Fields (1980) or Atkinson (1983).

11 Sen (1981) first used this approach, the entitlement approach, in his study of famines. A central point was that famines occur without a previous decline in food supply measured at the aggregate level. Studying the issues of access and relative availability was found to be central both for an explanation and prediction of the occurrence of famines.

12 In fact, Sen (1999) sees freedom as both the means and the end of development.

13 See, for instance, World Bank (1998).

14 Climate change originates from the total atmospheric concentration of greenhouse gases and the climate change damages of emitting one unit of GHG are similar irrespective of the location of the emission source.

REFERENCES

Atkinson, A B (1983) *The Economics of Inequality*, Oxford University Press

Atkinson, G and Pearce, D W (1993) 'Measuring Sustainable Development', *The Globe*, No 13, June, UK GER Office, Swindon

Baland, J and Platteau, J (1996) *Halting Degradation of Natural Resources: Is there a Role for Rural Communities?* Clarendon Press, Oxford

Banerjee, A, Besley, T and Guinnane, T W (1994) 'Thy Neighbour's Keeper: The Design of a Credit Cooperative with Theory and a Test', *Quarterly Journal of Economics*, Vol 109, pp491–515

Barbier, E (ed) (1997) 'The Environmental Kuznets Curve', Special Issue, *Environment and Development*, Vol 3

Beckerman, W (1994) 'Sustainable Development: Is it a Useful Concept?', *Environmental Values*, Vol 3, pp191–209

Bishop, R C (1978) 'Endangered Species and Uncertainty: The Economics of a Safe Minimum Standard', *American Journal of Agricultural Economics*, Vol 60, pp10–18

Bromley, D W and Feeny, D (eds) (1992) *Making the Commons Work: Theory, Practice and Policy*, ICS Press, San Francisco

Cantor, R and Yohe, G (2000) 'Economic Activity and Analysis' in Rayner, S and Malone, E L (eds) *Human Choice and Climate Change. The tools for Policy Analysis*, Vol Three, Batelle Press, Washington, DC

Cass, D (1966) 'Optimum Growth in an Aggregate Model of Capital Accumulation: A Turnpike Theorem', *Econometrica*, Vol 34, pp833–850

Ciriacy-Wantrup, S V (1952) *Resource Conservation: Economics and Policies*, University of California, Berkeley and Los Angeles

Coleman, J S (1990) *Foundations of Social Theory*, Harvard University Press, Cambridge, Mass

Daly, H E (1990) 'Toward Some Operational Principles of Sustainable Development', *Ecological Economics*, Vol 2, pp 1–6

Daly, H E (1991) 'Towards an Environmental Macroeconomics', *Land Economics*, Vol 67, No 2, pp255–259

Daly, H E (1995) 'On Wilfred Beckerman's Critique of Sustainable Development', *Environmental Values*, Vol 4, pp49–55

Dasgupta, P and Heal, G (1974) 'The Optimal Depletion of Exhaustible Resources' in *Review of Economic Studies Symposium: Economics of Exhaustible Resources*, pp3–28

Dasgupta, P (1993) *An Inquiry into Well-being and Destitution*, Clarendon Press, Oxford

Dasgupta P and Serageldin, I (1999) *Social Capital: A Multifaceted Perspective*, World Bank, Washington, DC

DENR (1997) *Department of Environment and Natural Resources: Procedural Manual for Department Administrative Order No 96–37*, Quezon City

Fields, G S (1980) *Poverty, Inequality, and Development*, Cambridge University Press, Cambridge

Grossman, M and Krueger, A B (1991) *Environmental Impacts of a North American Free Trade Agreement*, Working Paper No 3914, National Bureau of Economic Research, Cambridge, MA

Hanley, N D, Shogren, J F and White, B (1977) *Environmental Economics in Theory and Practice*, Macmillan Press, Basingstoke, UK

Hartwick, J M (1977) 'Intergenerational Equity and the Investing of Rents from Exhaustible Resources', *American Economic Review*, Vol 67, No 5, pp972–974

Hartwick, J M (1978) 'Investing Returns from Depleting Renewable Resource Stocks and Intergenerational Equity', *Economic Letters*, Vol 1, pp85–88

IUCN, WWF and UNEP (1980) *World Conservation Strategy*, International Union for the Conservation of Nature, Geneva, and UNEP, Nairobi

Koopmans, T C (1965) 'On the Concept of Optimal Economic Growth' in *The Econometric Approach to Development Planning*, North-Holland, Amsterdam

Malinvaud, E (1965) 'Croissances optimales dans un modele macroeconomique' in *The Econometric Approach to Development Planning*, North-Holland, Amsterdam

Narayan, D and Pritchett, L (1997) *Cents and Sociability: Household Income and Social capital in Rural Tanzania*, World Bank, Washington, DC

North, D (1990) *Institutions, Institutional Change and Economic Performance*, Cambridge University Press, New York

Olson, M (1982) *The Rise and Decline of Nations: Economic Growth, Stagflation and Social Rigidities*, Yale University Press, New Haven

Pearce, D W, Barbier, E W and Markandya, A (1990) *Sustainable Development*, Edward Elgar, London

Putnam, RD, Leonardi, R and Nanetti, RY (1993) *Making democracy work: Civic traditions in modern Italy*, Princeton: Princeton University Press

Rennings, K and Wiggering, H (1997) 'Steps Towards Indicators of Sustainable Development: Linking Economic and Ecological Concepts', *Ecological Economics*, Vol 20, No 1, pp25–36

Sen, A (1981) *Poverty and Famines: An Essay on Entitlement and Deprivation*, Clarendon Press, Oxford

Sen, A (1992) *Inequality Reexamined*, Clarendon Press, Oxford

Sen, A (1999) *Development as Freedom*, Alfred A Knopf, New York

Solow, R (1974) 'Intergenerational Equity and Renewable Resources', *Review of Economic Studies Symposium*, pp29–45

Solow, R (1991) *Sustainability: An Economist's Perspective*, Eighteenth J Seward Johnson Lecture

UNDP (1999) *Human Development Report*, Oxford University Press, New York

World Bank (1992), *World Development Report*, Oxford University Press, New York

World Bank (1997) *Expanding the Measure of Wealth: Indicators of Environmentally Sustainable Development*, Washington, DC

World Bank (1998) *World Development Indicators*, Washington, DC

World Commission on Environment and Development (1987) *Our Common Future*, Oxford University Press, London

Chapter 3

A Review of the Literature on Climate Change and Sustainable Development

Kirsten Halsnaes

INTRODUCTION

The purpose of this chapter is to assess the literature on sustainable development (SD) as a tool for, or an input in, climate change mitigation studies.[1] This research area is currently relatively new, and evolving, and the literature is therefore characterized by broad discussions about scope and approach. These discussions have also considered methodological and conceptual issues related to developing operational SD criteria for climate change mitigation analysis. A number of case studies have made an initial attempt to assess quantitative and qualitative SD impacts of climate change mitigation policies, and a number of results from these studies will be reviewed in this chapter.

Climate change mitigation studies have traditionally consisted of cost-effectiveness analysis of policy options based on various analytical approaches, such as global integrated assessment models, macroeconomic models, sectoral models, and technology and project assessment approaches. These studies have primarily assessed costs or welfare losses arising from climate change mitigation policies.

The introduction of sustainable development as a conceptual framework for climate change mitigation studies opens a number of new theoretical and conceptual issues as discussed in Chapter 2. The focus of this chapter is to review specific applications of the practical frameworks for sustainable development in the context of climate change mitigation studies that have been presented in the literature and to provide an overview of the main methodological similarities and differences.

Using SD as a framework for assessing climate change mitigation policies introduces a new and broader set of policy objectives in order to reflect development, equity and short- and long-term sustainability aspects. Following this, it also becomes relevant to examine policy options that, in addition to climate change mitigation, are designed to contribute to the broader sustainable development policy objectives.

These broader policy objectives that reflect SD aspects have been addressed in a number of studies that have typically included local environmental policy objectives related to air and water quality. This work has drawn on the rich existing literature on environmental externalities, including theoretical and empirical studies on various global, regional and local environmental problems. Much of it has been carried out around technical discussions that have been conducted as a side-activity to the ongoing Third Assessment of the Intergovernmental Panel on Climate Change (IPCC). The same issues have also be analysed in discussions about the implementation of the Clean Development Mechanism (CDM) of the Kyoto Protocol (UNFCCC, 1997). As noted in Chapter 1, CDM is a project-based mechanism that facilitates cooperative greenhouse gas (GHG) emission-reduction policies between countries that have a reduction commitment according to the Kyoto Protocol and countries without such a commitment. Article 12 of the Kyoto Protocol states that CDM projects have to 'assist sustainable development' in the project host countries. The CDM-related papers on SD typically includes a general discussion about how SD can and should be interpreted in relation to the mechanism and suggests a number of SD impact indicators. A number of the papers also include project case studies, where SD impacts of CDM projects are assessed.

This chapter reviews a number of conceptual and methodological studies on SD applied as a framework for climate change mitigation studies. It compares the SD indicators that have been suggested in the literature, and continues with a comparative assessment of the quantitative and qualitative results.

THE FRAMEWORK FOR ASSESSING SUSTAINABLE DEVELOPMENT IMPACTS OF CLIMATE CHANGE MITIGATION

Introduction

This section includes an overview of conceptual framework for climate change mitigation analysis within the paradigm of SD. It distinguishes between approaches that take as their starting point the general SD literature, and those that primarily are an extension of the analytical methods that traditionally have been applied to climate change mitigation costing studies.

The models for the assessment of SD vary from very general approaches that focus at the global policy level, to approaches that focus on regional and national impacts of specific projects and policies. Given the generally weak theoretical basis for SD, it is not surprising that many analytical approaches have emerged at each level of analysis, and that is indeed the case. One thing that most models have in common, however, is that they are by nature multidisciplinary. This implies that a broad range of different analytical tools is needed to carry out the assessment, including ecological modelling, welfare economics, technology assessment and social impact assessment.[2]

Overview of the Framework for SD Assessment

As noted above, the literature on SD and climate change has two important sources: the IPCC-related literature and the CDM-related literature. This section reviews the models that have been suggested in relation to IPCC discussions and the operationalization of the CDM mechanism.

Conceptual Framework for Assessing SD as a Broad Policy Objective of Climate Change Mitigation Policies

Working Groups II and III (in particular III) of IPCC (Metz et al, 2001) have initiated, as part of the Third Assessment process, a number of technical expert meetings and cross-cutting discussions regarding the development of a framework for assessing climate change policies in the context of sustainable development. A number of these discussions are reported by Banuri et al (2001)[3] and Munasinghe (2000).

Banuri et al (2001, p93) describe the overall objective of the conceptual work on SD as a need to develop a framework that can be used to compare the policy options that emanate from concerns about global equity. It is stated that the framework has evolved to enable the assessment of the synergies and trade-offs involved in the pursuit of multiple goals – environmental conservation, social equity, economic growth and poverty eradication. The framework integrates the conceptual basis of three domains – namely, economic, environmental and social.

Sustainability is presented by Banuri et al as an integration of the economic, environmental and social assessment, and the main conceptual drivers of these three domains are presented as efficiency, equity and sustainability. The label efficiency is used to describe 'conventional' climate change mitigation studies that have focused on the assessment of cost-effective policies and instruments given conditions of equity and sustainability, but not primarily guided by them. Equity is introduced in the policy framework to allow the inclusion of considerations that are not primarily motivated by climate change, but rather to examine the impact of climate change and mitigation policies on existing inequalities. It is also suggested to use climate change policies as an instrument to promote equity.

The sustainability aspect is introduced in the analysis through emphasis on alternative long-term development pathways. This level of analysis could consider, for example, development objectives of nations, intra- and intergenerational equity and long-term institutional changes and policy options.

Turning to the assessment of mitigation options, Banuri et al, pp95–98, suggest that the sustainable development approach can be applied by conceptualizing climate as a kind of capital. Since the prospects of human welfare and income – given technological and financial conditions – depend on the right to emit carbon, a restriction on this right means a diminishment of development prospects, unless it is compensated by increases in other forms of capital: financial, technological, social human. In the short run, such loss can be compensated (incompletely) only by additional financial resources or access to clean technologies. In the long run, the policy options for sustainable development include options that improve the capacity of decision-making, including the capacity of entrepreneurs, investors, educators, students and policy-makers to respond to climate change.

Banuri et al outline, on a very preliminary basis, a framework for assessing the efficiency, equity and sustainability dimensions of climate change mitigation policies, which is labelled 'mitigative capacity'. Mitigative capacity is seen as one critical component of a country's ability to respond to the mitigation challenge. They conclude that development, equity and sustainability objectives, as well as past and future development trajectories, play critical roles in determining which alternative mitigation option should be adopted and when it should be adopted. Following that, it is expected that policies designed to pursue development, equity and/or sustainability objectives might be the most effective climate change mitigation policies. The final conclusion is that, due to the inherit uncertainties involved in climate change policies, enhancing mitigative capacity can be a policy objective in itself (Banuri et al, 2001, pp103 and 108).

The focus on mitigative capacity as the main objective in assessing climate change mitigation policies opens the range of potential policy options to include various general social and development policies. The critical issue is then to assess the opportunity cost of using such 'broad' policy options to address specific climate change issues compared with other options that are designed and targeted more directly to address the particular social and developmental issues. The dilemma between broad SD-oriented policies and specific policies targeted to address different aspects of climate change, other environmental policies and social priorities is further discussed later in this section in relation to the framework for assessing the ancillary benefits of climate change mitigation projects.

The IPCC TAR has included, as mentioned earlier, a number of technical expert meetings that have discussed specific cross-cutting issues related to emission and stabilization scenarios, cost concepts, ancillary benefits, decision-making frameworks, uncertainty and sustainable development. Munasinghe (2000) has developed a cross-cutting issue paper on SD as part of this background work.

The framework for assessing the SD impacts of climate change mitigation policies is presented in the following way by Munasinghe. He says:

> *From the operational viewpoint, so-called 'win-win' climate change strategies are the most desirable – ie those that enhance all three elements of sustainable development (economic, social and environmental). Policies and measures which advance one element at the expense of another need to be analysed within a framework that facilitates trade-offs (eg increase manufactured capital while depleting both social and natural capital: or improve the resilience of a social system while increasing the vulnerability of an ecosystem) . . . When all important impacts of a specific climate change option may be valued in economic terms, the usual approach of comparing the corresponding costs and benefits will provide useful insights. Where certain critical impacts cannot be valued (ie reduced to a single monetary 'numeraire') other techniques such as multi-criteria analysis could be helpful. High levels of uncertainty and risk might be dealt with through the use of modern decision analysis frameworks. (pp 85–86)*

Munasinghe's framework emphasizes the assessment of SD impacts of climate change mitigation. The SD impacts that are considered include a broader range of policy objectives than those contained in conventional cost-effectiveness analysis, and a number of specific constraints are defined in relation to economic, environmental and social dimensions of sustainability. It is recommended that an integrated assessment of economic, social and environmental impacts is conducted, and that consideration is given to potential trade-offs between the three impacts areas through decision analysis approaches that address resilience issues and high levels of uncertainty and risk. A conceptual framework for the integrated assessment of efficiency, resilience issues, uncertainty and risk is outlined by Van Pelt, 1993. It is suggested here that an assessment of efficiency should be expanded with criteria such as ecological sustainability, where the criteria should be considered as a constraint in relation to a given threshold of ecological resources such as biodiversity, non-renewable resources, and annual regeneration rates for renewable resources.

Following that, it can be concluded that Munasinghe's framework can be seen primarily as a further development of conventional policy effectiveness analysis, where climate change mitigation is the main policy objective guiding the selection of policy options, but the assessment of policies is based on a number of restrictions related to equity and ecological resilience.

Conceptual Framework for Assessing Ancillary Benefits

A framework for assessing the ancillary benefits of climate change mitigation policies has been suggested based on an expansion of conventional cost-effectiveness studies and cost-benefit analysis approaches (OECD, 2000). The basis for this work has both been climate change mitigation costing studies and broader environmental externality studies, considering local air pollution, water pollution, waste disposal, and so on.

Krupnick, Burtraw and Markandya (2000) define ancillary benefits as feedbacks on GHG-emission reduction policies into the economic system in the form of changes in other pollutants and non-environmental externalities (such as changes in employment). These feedbacks are termed externalities to emphasize that the focus of assessing ancillary benefits is on cases where a policy yields a change in the productive use of resources or in the welfare of individuals that are not fully taken into account by the agents involved.

The IPCC TAR, Chapter 7, has similarly concluded that 'ancillary benefits' are the ancillary effects of climate change mitigation policies on problems other than GHG emissions (Markandya et al (2000) pp160–161). These effects include, for example, reductions in local and regional air pollution, associated with the reduction of fossil fuel use, and indirect effects on issues such as transportation, agriculture, land use practices, employment and fuel security. In parallel to that, the IPCC report introduces the term 'co-benefits' that by definition address the same side-impacts to GHG emission-reduction policies as 'ancillary benefits' but treat these impacts as joint policy objectives to climate change mitigation rather than as indirect impacts to the policies.

The focus of the ancillary benefit literature has been on the short-term impacts of climate change mitigation, giving main attention to the inclusion of health impact estimates that are considered to be a dominant side-impact related to local air pollution. While the benefits of climate change mitigation is a global intergenerational common good, ancillary benefits related to local air pollution largely accrue in the present and in an institutional context that is largely commensurate with policy-making (at the national level).

Returning to the discussion about a sustainable development framework for assessing climate change mitigation options, it can be said that the ancillary benefit concept primarily is an extension of traditional efficiency analysis of environmental policies with a number of additional policy objectives related to expected side-impacts of the options. This line of development is further expanded in relation to the above-mentioned 'co-benefit concept'. The ancillary benefit concept, as well as the co-benefit concept, focuses on efficient economic and natural resource management seen in the short-term perspective, while excluding the long-term aspects related to social impacts, equity and other sustainability issues.

A number of critical methodological issues arise in the application of the ancillary benefit framework to empirical studies. Krupnick, Burtraw and Markandya (2000) have highlighted the following issues:

• The estimation of ancillary benefits requires localized models of environmental impacts, population, exposure, preferences and valuation that should be based on a consistent methodology. The result is expected to be estimates that vary significantly by nation and region, and with time.
• Some climate change mitigation policies can lead to ancillary costs rather than benefits.
• The ancillary benefits depend on assumptions about policy objectives related to the externalities considered. These assumptions should be reflected in the baseline case of the assessment and should include time-dependent assumptions about regulation of the specific side-impacts included.
• The prevalence of externalities relies on the assumptions that these are not already internalized, for example, through insurance systems, risk premiums in salaries and other sorts of compensations. This implies also that the existence of externalities presumes that a sort of market barrier precludes their internalization.
• A number of specific externalities are particularly significant in developing countries and will therefore often be relevant to include in the analysis. These include unemployment and equity issues.
• The ancillary benefits or co-benefits of climate mitigation policies must be assessed in relation to the opportunity costs of pursuing the different policy objectives by more dedicated options that address the specific issues. Such a dedicated option – for example, in the case of local air pollution – can include various flue gas cleaning technologies.

Markandya (1998) has outlined an operational framework for assessing indirect costs and benefits of GHG emission-reduction projects that is an example of an application of the ancillary benefit concepts. The basic idea is to extend the assessment of the direct costs of GHG emission-reduction projects which have been made in various national climate change mitigation studies with additional indirect cost and benefit indicators. The suggested indicators include: employment impacts, income gains and losses of different groups, environmental changes related to local externalities and sustainability indicators. These indicators can be supplemented further with indicators that reflect other macroeconomic impacts, poverty alleviation and other social issues. The methodological approach for measuring and comparing the direct as well as indirect policy impacts is to use a mixture of monetary values, other physical indicators and qualitative information (Halsnaes, Callaway and Meyer (1998); Markandya (1998).

Sustainable Development Related to the Clean Development Mechanism of the Kyoto Protocol

The CDM mechanism of the Kyoto Protocol includes language stating that CDM projects should assist sustainable development in the host countries. A number of international studies have therefore made a first attempt at developing a framework for assessing SD impacts of CDM projects. Two of these (for example, Brazil and India) are included in this book in Chapter 8.

The major body of the international literature has suggested that SD impact indicators should be based on national development priorities in project host countries and that the SD impacts of CDM projects more specifically should be assessed in relation to the specific site and institutional context of the projects, and this has been the approach of all papers included in this review.

Austin et al (2000) define SD impacts as co-benefits in the CDM host country related to air and water pollution, other environmental impacts, biodiversity and social impacts, including employment generation and income distribution impacts. The framework is only set out in broad terms and has therefore been operationalized somewhat differently in case studies for Brazil, China and India.

The Brazilian study has defined SD impact indicators on the basis of suggestions from local policy-makers. The suggested indicators then are supposed to be used in a broad decision-making process where the qualitative information is used to develop a national strategy. The Chinese approach for selecting SD impact indicators is linked to official national environmental policies on air and water pollution control, and the co-benefits of the CDM projects on these areas are assessed as a background for national priority-making. The study for India, similarly to the Chinese study, takes the starting point in national development plans that both include economic, social and environmental development goals. The CDM projects are in addition to the GHG emission-reduction potential evaluated against three groups of national Indian priorities – namely, development, non-climate environmental benefits and feasibility in implementation. Chapter 8 provides more details of the Brazilian and Indian studies.

A review of papers on the SD impacts of CDM projects leads to the conclusion that there are many similarities across papers both in analytical approach and in specific indicators suggested (based on Munasinghe (2000); James and Spalding-Fecher (1999), Serôa Da Motta et al (2000) and Iyer (1999)). The SD impact indicators are in these studies grouped into economic, environmental, and social impacts, where a few of the studies also include impacts on technological development.

The studies are very similar in their suggestions regarding economic impact indicators. The main difference is here seen between studies that primarily focus on expected project level impacts, such as employment generation, and others that include macroeconomic impacts, such as macroeconomic stability, balance of payment and regional economic development (Thorne and La Rovere (1999) and Serôa da Motta et al (2000)). Social impacts are also treated similarly in the studies, where employment generation and income distribution issues are included in the indicator list in most studies. Other social impact indicators suggested are poverty alleviation, cultural/heritage aspects and consultation/enpowerment, which are related to social capital aspects in Munasinghe's paper (Munasinghe, 2000).

Environmental indicators include local and regional air pollution, water quality and availability, flood protection, soil conservation, waste, noise, bio-diversity and sustainability indicators for natural capital.

The indicators related to technological development are treated somehow differently in the studies. The Iyer (1999) study considers technical capability and learning curve aspects as part of the social CDM project impacts, while others suggest to include a more separate technology-oriented indicator like 'access to state-of-the-art technology'. The James and Spalding-Fecher approach is more focused on the feasibility aspects of the technologies seen in relation to existing national technical systems and support.

An overview of the indicators suggested by the different studies is given in Table 3.1. We can compare these indicators with those derived in Chapter 2 on the basis of more theoretical sustainability considerations. The following points of similarity and difference emerge:

1   The environmental indicators in this table include many of those in Table 2.2. The latter, however, define some of the measures in more detail – for example, the contribution of the project to the cost of renewable and non-renewable energy, now and in the future; the impact on the move to renewable energy use overall; and the use of renewable and non-renewable non-energy materials. If successful, these indicators can only strengthen the environmental indicators listed in Table 3.1.

2   The 'economic' and 'social' indicators in Table 3.1 overlap with those in Table 2.3, with considerable emphasis in each case on employment effects, inequality and regional multipliers. Table 2.3 also includes a measure of genuine savings and of food supply. Table 3.1, on the other hand, looks at 'process' indicators – ie, those that do not bear directly on welfare but may point to impending changes in welfare, such as the trade balance, inflation, competitiveness, and so on.

3   Table 3.1 has no discussion of social acceptability or social capital as such, with the exception of Munasinghe (2000), who notes the importance of consultation and empowerment. Tables 2.4 and 2.5 provide a much richer list of 'filter' type indicators which, as noted in Chapter 2, define the reference operating conditions that exist prior to the project and that should be satisfied if the project can be undertaken with confidence in the first place.

4   Table 3.2 includes one set of indicators that do not appear separately in Chapter 2 and these relate to technological development. Partly these are

included in Table 2.3 (under economic indicators), where sustainable use of natural resources is mentioned. Others, such as 'access to state-of-the-art technology' and 'availability of technological support' may be useful in specific conditions in comparing projects. We have not seen, however, any applications of these in an actual appraisal.

From this it can be concluded that SD indicators for CDM projects in the studies reviewed can be characterized as an expansion of existing approaches for climate change mitigation studies with more policy objectives and constraints, and not similar to those derived from an SD perspective for climate change in general. The SD criteria actually selected will be based on national development object-ives, which cover a broad range of economic, social and environmental issues that are represented by relatively similar indicators in the studies reviewed.

Conclusion on the Conceptual Frameworks

The development of a Sustainable Development Framework for assessing climate change mitigation projects has been initiated by several authors with the aim of establishing operational decision criteria for integrating SD objectives into climate change policies. The conceptual work in this field has included global climate policies from both a short- and long-term perspective and more specific issues related to international climate change finance, as, for example, in relation to the CDM mechanism.

Some of the approaches suggested are primarily based on the rich theoretical literature on sustainable development and consider the various aspects of devel-opment, environment and social dimensions of this problem. In this they have a lot in common with the discussion in Chapter 2. These approaches recommend addressing SD issues through an integrated multidisciplinary analytical approach. The focus is on assessing a country's mitigative capacity, which is seen as one critical component in a country's ability to respond to the mitigation challenge. It is concluded that a country's development, equity and sustainability objectives, as well as past and future development trajectories, are critical to its mitigative capacity, and policy options basically therefore should address all these aspects of development, equity and sustainability.

Other approaches, including the Ancillary Benefit Approach, can be seen primarily as further developments of conventional cost-effectiveness and CBA analysis that have been used in climate change mitigation studies, as well as in other environmental assessments. The scope of these approaches is to include a number of side-impacts to climate change mitigation policies. The approaches emphasize a focus on short-term environmental externalities which in some cases have been supplemented with social impact indicators related to employment generation and/or income distributional impacts. Here the link to the SD liter-ature is less strong, although there is still a great deal of overlap in practice.

In the next section we look at a number of empirical studies, both in the Ancillary Benefit framework and in the broader CDM framework, and comment on their internal consistency and usefulness, as well as their comparative strengths and weaknesses.

Table 3.1 Suggested SD indicators in the studies

Approach	Economic indicators	Environmental indicators	Social indicators	Technology development indicators
Munasinghe (2000) Mixture of monetary, quantitative and qualitative information including stock variables and assumptions about non-substitutable resources	– Growth – Efficiency – Stability	– Growth – Efficiency – Stability	– Poverty – Culture/heritage – Consultation/empowerment	
Markandya (1998) Monetary values supplemented with quantitative and qualitative information	Employment impacts Adjustment to reflect: – The value of improved health conditions by employment – A low shadow price of presently unemployed labour – The value of lost leisure time – Employment	Associated environmental changes Adjustment to reflect: – local and regional air pollution damages in monetary values – Sustainability indicators in qualitative terms related to natural capital	Income gains and losses of different groups Adjustment to reflect: – Income distribution weights	
Austin et al (2000) MCA framework assigning low-, medium- and high-impact indicators to the indicators		– Air quality – Water quality – Water availability – Soil conservation – Solid waste – Noise – Flood prevention/protection – Biodiversity protection	– Rural development – Poverty alleviation and equity	
Thorne and La Rovere (1999) MCA that with specific scoring system for each indicator	– Balance of payments sustainability – Macroeconomic sustainability	– Local environmental sustainability (change in level of selected pollutants)	– Contribution to net employment generation	– Technological self-reliance (foreign exchange)

Equal weights are applied to all indicators	– Cost-effectiveness (project evaluation)	– Local environmental impacts	– Employment generation	– Sustainable use of natural resources (renewable energy resources) – Access to state-of-the-art technology
Banuri and Gupta (2000) Qualitative ranking system with high, medium, and low values	– Financial additionality to exclude negative cost options			
James and Spalding-Fecher (1999) MCA with formalized indexes and weighting systems for each indicator	– Macroeconomic – Balance of trade – GDP – Inflation – Return on initial investment – International competitiveness	– Soil conservation and biodiversity – Water resources and biodiversity – Air quality, non-GHG emissions – Leakage	– Social equity and poverty alleviation – Job creation – Sectoral and institutional capacity for implementation – Administration burden and barriers	– Availability of technological support – Compatibility existing technical systems – Technology and associated inputs
Seróa de Motta et al (2000) MCA where the impacts are input to a qualitative assessment using a high-, medium-, and low-value scoring system	– Aggregate demand – Trade balance – Regional economy – Opportunity cost of output forgone	– Water resources – Water resource quality – Urban air pollution – Soil erosion – Biodiversity protection	– Income distribution impact based on the project's unskilled labour participation – Consumption for different income classes – Environmental benefits for different income classes	
Iyer (1999) MCA approach		– Local environmental and health impacts – Acidification – Air quality – Waste disposal	– Employment generation	– Capability building (technical skills for repair and maintenance) – Enhancing the technological base (shortening the learning curve)

Note: The same indicator is included in some studies as an economic indicator and in other studies as a social indicator. This is, for example, the case for the indicator 'employment generation'

Empirical Study Results

Introduction

As this subject area is relatively new, few studies have been published so far and these primarily are a further development of conventional climate change mitigation studies based on cost-effectiveness analysis or multicriteria assessments. The next section looks at ancillary benefit studies, followed by a section on CDM studies.

Ancillary Benefit Studies

Several studies have assessed the ancillary benefits of climate change mitigation policies with a particular weight on health impacts of reduced SO_2, NO_x and particulate emissions. For the USA and European countries, the studies have been closely related to national or regional air pollution studies (Rowe et al, 1996; ExterneE, 1995; ExterneE, 1997). A number of studies are also available for developing countries, including Brazil, Chile, China and South Korea.

Cross-country comparisons of ancillary benefit studies are difficult to conduct for various reasons. The studies are not methodologically consistent, in particular regarding assumptions on policy assumptions, and they do not include the same externalities. Another major difference across studies is the climate change mitigation policy objectives considered. Some studies cover relatively small GHG emission-reduction targets and the ancillary benefits of these, and others cover very large reduction targets.[4]

One way to measure the ancillary benefits of GHG emission-reduction policies is to assess the benefit on other policy objectives like SO_2, NO_x and particulates per unit of GHG emission reduction. The GHG abatement cost per tonne of carbon in this way can be compared with the ancillary benefit per tonne of carbon, IPCC, 2001 (Chapter 8, Table 8-2.4A) includes a review of GHG abatement costs and ancillary benefit estimates in studies for Chile, China, Hungary, Norway, Western European countries and the US.[5]

The IPCC review concludes that in the case of Chile, China and Hungary the ancillary benefits of climate change mitigation policies considered exceeds the abatement costs, with the implication that the 'net' cost of the carbon reduction is negative. The Western European studies find that abatement costs and ancillary benefits are about the same level, while the Norwegian study and studies for the US find that ancillary benefits are lower that GHG abatement costs. It is also concluded that the magnitude of the ancillary benefits seems to be lower for the US and Western European countries than for the developing countries being considered. This result therefore reflects that the US and the Western European countries in the last decades already have implemented control policies for the air pollutants and other externalities considered, implying that the intensity of these externalities per unit of GHG emission reduction is relatively low.

O'Connor (2000) explains this difference between ancillary benefit estimates for developing countries and industrialized countries in relation to a Kuznets curve argument, arguing that industrialized countries have moved further out of their inverted U-shaped curve for control of local air pollutants. This suggests an hypothesis that the lower a country's level of development, the larger are the ancillary benefits of GHG emission-reduction policies.

The study review shows that there are also major differences between study results for the same country. This is seen for the US and many Western European countries where several studies have been conducted (Burtraw et al, 1999). Differences also arise for different studies that have been conducted for Chile and China. A specific uncertainty which applies to these studies is that they have been based on different assumptions about the valuation of health damages using a Value of Statistical Life (VOSL) estimates. Major differences have arisen in the case of Chile because studies have based their VOSL estimates on different per capita gross domestic product (GDP) estimates (O'Connor, 2000).

These differences in study results reflect the inherent complexities in applying a consistent methodological framework to ancillary benefit studies. An overview of the detailed study results is included in Table 3.2.

Markandya's framework for assessing ancillary benefits has been applied in a number of case studies for developing countries and countries with economies in transition, and has included project co-benefits related to local air pollution, employment generation and income distribution (Markandya, 1998; Halsnaes and Markandya, 1999). The case examples include a number of renewable energy projects such as wind turbines, solar water heating, PVs, biogas and bagasse power production. Another category is end-use efficiency improvement projects, including industrial motors and energy savings in buildings. This is supplemented with a forestry case example from Russia and liquefied petroleum gas (LPG) buses in Mauritius. These examples are not at all comprehensive in the coverage of potential mitigation options or national variety, but rather are illustrative examples of what can be included in a broader assessment of direct as well as indirect project impacts.

An overview of the abatement costs and ancillary benefits of the case projects is given in Table 3.3. From this one can conclude that the application of the framework for assessing the abatement costs and ancillary benefits to the ten case examples has major implications on GHG abatement costs. In nine out of the ten case example the projects have been assessed to have net ancillary benefits related to employment generation, local air pollution reduction and/or income distribution benefits. One project had a net ancillary cost because it implied decreased employment. Most of the identified ancillary benefits were assessed to be very site and context specific in depending on local labour market conditions, income distribution and air pollution. The ancillary benefits were, except in the case of Hungary, assessed to be relatively small compared with the abatement costs. This was the case in particular for the GHG emission-reduction options with relatively high abatement costs, such as some of the options for Mauritius. It must be recognized that this result cannot be compared directly with the conclusions of the IPCC review, which found larger ancillary benefits compared with abatement costs, because this conclusion was the outcome of more general cost-effectiveness studies of options which were assessed to have low abatement costs.

CDM Studies

The Austin et al study (2000) includes a comparative assessment of case studies for Brazil, China and India and discusses the cost-effectiveness of different types of CDM projects in meeting different policy objectives, such as donor country

Table 3.2 *Scenarios and studies reviewed by IPCC, 2000*

Study	Area and sectors	Scenarios 1996, US$	Average ancillary benefit US$ per t C	Key pollutants	Major endpoints
Dessus and O'Connor (1999)	Chile (benefits in Santiago only)	Tax of $67 (10% C reduction) Tax of $157 (20% C reduction) Tax of $284 (30% C reduction)	$251 $254 $267	7 air pollutants	Health – morbidity and mortality, IQ
Cifuentes et al (2000)	Santiago, Chile	Energy efficiency	$62	SO_2, NOx, CO, NMHC. Indirect estimations for PM_{10} and resuspended dust	Health
Garbaccio, Ho and Jorgenson (2000)	China – 29 sectors (4 energy)	Tax of 1$ per t C Tax of 2$ per t C	$52 $52	PM_{10}, SO_2	Health
Wang and Smith (1999)	China – power and household sector	Supply-side energy efficiency improvement. Least cost per unit GWP reduction fuel substitution		PM, SO_2	Health
Aunan, Aaheim, Seip (2000)	Hungary	Energy conservation programme	$508	TSP, SO_2, NO_x, CO, VOC, CO_2, CH_4, N_2O.	Health effects, material damage vegetation damage

Study	Country	Value	Tax	Pollutants	Effects
Brendemoen and Vennemo (1994)	Norway	$246	Tax $840 per t C	SO_2, NO_x, CO, VOC, CO_2, CH_4, N_2O, particulates	Indirect: Health costs, lost recreational value from lakes and forests, corrosion. Direct: traffic noise, road maintenance, congestion, accidents
Barker and Rosendahl (2000)	Western Europe (19 regions)	$153	Tax $161 per t C	SO_2, NOx, PM_{10}	Human and animal health and welfare, materials, buildings and other physical capital, vegetation
Scherega and Leary (1993)	US	$41	Tax $144 per t C	TSP, PM_{10}, So_x, NO_x, CO, VOC, CO_2, Pb	Health – morbidity and mortality
Boyd, Krutilla, Viscusi, (1995)	US	$40	Tax $9 per t C	Pb, PM, SOx, SO_4, O_3	Health, visibility
Abt (1999)	US	$8 $68	Tax $30 per t C Tax $67 per t C	Criteria	Health – mortality pollutants and illness, visibility and household soiling (materials damage)
Burtraw et al (1999)	US	$3 $2 $2	Tax $10 per t C Tax $25 per t C Tax $50 per t C	SO_2, NO_x	Health

Table 3.3 *Abatement costs and ancillary benefits of case examples from developing countries and economies in transition*

	Abatement cost US$ per t CO$_2$	Local air pollution benefit US$ per t CO$_2$	Employment benefit US$ per t CO$_2$	Income distribution benefit US$ per t CO$_2$
Biogas plant, Tanzania	20.9	1.4	1.7	51.4
Foresty project, Russia	10.3	–	2.1	–
Industrial motors, Thailand	–0.1	0.02	–	–
Solar water-heaters, Hungary	179.5	44.2	88.7	–
Window insulation, Hungary	20.4	12.2	5.3	
Wind turbines, Mauritius	113.7	2.4	10.8	–
LPG buses, Mauritius	1195.5	20.4	5.4	–
Solar water, Mauritius	136.5	18.1	22.3	–
Bagasse, Mauritius	807.2	48.2	–69.9	–
PV light, Mauritius	6964.7	24.1	689.1	–

Source: Halsnaes and Markandya, 1999

objectives, to minimize financial costs and host country objectives related to potential side-impacts on SD. The Brazil and India studies are discussed in greater depth in Chapter 8.

The country studies were conducted by national experts and included an assessment of national development issues and priorities to be used as a basis for selecting SD impact indicators. Existing GHG abatement studies for the countries were reviewed and a number of the policy options included were selected for a more detailed assessment of financial project costs and SD impacts. The trade-offs and synergies between GHG emission-reduction benefits and SD benefits of the projects were finally assessed. Each national study followed the same analytical steps but exhibited major differences in analytical approaches and selected SD impact indicators.

The comparative assessment of the country studies concluded that many of the CDM projects reviewed were assessed to have large co-benefits on SD indicators in addition to offering low GHG abatement costs. The SD impact indicators that were assessed to have high scores were those of improved water and air quality, employment creation as well as local energy sufficiency. A number of negative SD impacts of CDM projects were also identified. In India these included increases in waste from some energy technologies, and in Brazil they arose from negative impacts on soil and water quality from the use of chemicals in plantations.

The Brazilian study included a number of forestry and energy sector projects, which are shown in Table 3.4 (Serôa da Motta, Young and Ferraz, 2000).

Table 3.4 *GHG abatement costs of forestry and energy sector projects in Brazil*

	Abatement cost with 12% discount rate US$ per t of C
Forestry projects	
Pulp plantation in degraded area	1.4
Charcoal plantation in degraded area	0.7
Saw-logging plantation in degraded area	−9.5
Private sustainable native forest management for saw-logging	9.0
Public concession forests for saw-logging	1.8
Concession forest as part of a conservation programme	5.0
Energy projects	
Ethanol with bagasse cogeneration	19.7
Cogeneration from refineries	−17.7
Biomass, thermoelectric gasification of wood	2.4
Wind energy	14.6

Source: Serôa da Motta, Young and Ferraz, 2000

The abatement costs of the forestry and energy sector projects as reported in the table reflect the financial cost perspective by the donor countries. It can be seen that the most attractive forestry options regarding this perspective are saw-logging plantations, charcoal plantations, pulp plantations and public concession forests for saw-logging. The most attractive CDM investment projects in the energy sector in the same perspective are cogeneration from refineries and biomass thermoelectric gasification of wood.

The impacts on SD of the potential forestry and energy sector projects have been qualitative, assessed in relation to 12 indicators that reflect impacts on the environment, development and equity (see a detailed overview of the indicators in Table 3.1). The qualitative assessment of the SD impact indicators concludes that the SD impacts of most projects are either positive or neutral. For the forestry options in particular there are found to be a large number of positive SD impacts of the private sustainable native forest management project and the concession forest project (that are the projects with the highest abatement costs), while the SD impacts of the other projects are low or neutral.

On the energy side, cogeneration from refineries is assessed to be attractive because it has neutral or positive impacts on the selected SD impact indicators and has low abatement costs. In the same way wind turbines only have neutral or positive SD impacts, but are not considered to be very attractive due to relatively high abatement costs. Biomass electricity as a substitute to charcoal has secondary benefits related to economic development and equity, but is assessed to have low negative impacts on water access and quality, urban air pollution and soil erosion. The ethanol option, which is a relatively expensive abatement option, is expected to have high development and equity benefits, but eventually will have a small negative environmental impact on water and soil erosion.

On this basis it is concluded that the CDM projects in Brazil that seem to be the most attractive from the perspective of domestic development, environmental and equity considerations are not the same as the ones that international investors will pick up as those that offer the highest return.

The CDM study for India included an evaluation of forestry and energy sector options (Pathak, Srivastava and Sharma, 1999). The abatement costs of the options were assessed and the energy sector projects then were ranked according to development criteria reflecting the feasibility in project implementation, non-climate environmental benefits, and the *development benefits* of the projects (see a detailed overview of these indicators in Table 3.5).

The project ranking was done with an analytical hierarchical process (AHP) technique where the development criteria and the cost-effectiveness were ranked. This procedure involved the assignment of weights to the four criteria and giving scores to the projects according to whether they were preferred or just equal to others. The two criteria with the highest weight were feasibility and development (each with 39 per cent out of 100 per cent total weighting). Other environmental benefits were given a lower weight of about 14 per cent while cost-effectiveness was only given a weight of about 8 per cent.

It was found that there was an overlap between the two projects with the highest development benefit ranking and the two options with the lowest abatement cost for the subsectors conventional power generation, renewables for power generation, renewables for agriculture, and cement, iron and steel manufacture. There is some difference between the development benefit ranking and the abatement cost ranking for projects with higher costs, but generally it can be concluded that the Indian study identified a lot of synergies between abatement cost perspectives and SD priorities. The same was concluded for the qualitative assessment of the forestry sector projects. An overview of the projects and their abatement costs together with information about the ranking of the projects according to development criteria is given in Table 3.5.

The CDM study for China included the energy sector and one forestry project (Ji and Jungfeng, 2000). The GHG abatement costs of the projects were assessed and supplemented with a general qualitative assessment of the development impacts related to environmental policies and economic development. An overview of the projects is given in Table 3.6.

The qualitative assessment of development benefits for China concludes that there are very large potential development benefits of CDM projects, mainly arising from the co-benefits of decreased local air pollution. This is the case particularly when CDM projects substitute coal consumption or increase the

Table 3.5 *GHG abatement costs and ranking of energy sector and forestry projects in India*

	Abatement costs US$ per t C	Ranking by overall development benefits
Conventional power generation		
Bagasse-based cogeneration	−244	1
Combined cycle generation (natural gas)	−133	2
Atmospheric fluidized bed combustion	7	5
Pressurized fluidized bed combustion	47	4
Pulverized coal super-critical boilers	96	6
Integrated gasification combined cycle	96	3
Renewable for power generation		
Small hydro	29	2
Biomass power	134	1
Wind farm	216	3
Photovoltaic	1306	4
Renewables for agriculture		
Wood-waste gasifier	169	1
Agro-waste gasifier	177	2
Wind well (shallow)	298	5
Wind well (deep)	329	4
Photovoltaic pump	6333	3
Cement, iron and steel manufacture		
Dry suspension preheater kiln	7	1
Dry precalciner kiln	214	2
Forestry		
Natural regeneration	6	–
Enhanced regeneration		
Marginally degraded level	33	–
Totally degraded land	40	–
Private land (agroforestry)		–

Note: No formal ranking procedure was carried out for the forestry sector projects

efficiency of present use. Other positive SD impacts are assessed to arise from CDM projects that imply improved industrial competitiveness, rural and off-grid electrification, wastewater treatment, mining safety, soil retention, flood prevention and local employment.

One of the general conclusions of the study for China is that although carbon gains are insensitive to location, a reduction in air pollution or the creation of employment opportunities will depend on the site of the abatement projects.

General Conclusions on Empirical Studies

A number of empirical studies have assessed side-impacts of climate change mitigation policies, including impacts on the local environment, social impacts and broader development impacts. One class of studies – the ancillary benefit

Table 3. 6 *GHG abatement costs for energy sector and forestry projects in China*

	Abatement costs US$ per t C
Conventional power generation	
Fuel switching	25–100
Clean coal technologies	67
Coal-bed methane	–4
Biomass gasification	30
Industrial sector	
Anaerobic wastewater treatment and gasification	11
Equipment efficiency	< 1–25
Renewables	
Wind power	15
Solar power	
Solar thermal applications	6
Solar PV	60
Geothermal	14
Afforestation	
Afforestation	–13–2

studies – have emphasized the health impacts of environmental policies, while the other set includes recent CDM project studies which have assessed side-impacts in relation to a range of SD impact indicators in addition to indicators reflecting the local environment.

A general conclusion of these empirical studies is that there are significant side-impacts from most of the climate change mitigation policies assessed. The ancillary benefit studies have concluded that the ancillary benefits per unit of carbon reduction for some options sometimes can be as large as the abatement costs, but this result is very site and context specific. Studies that are assessing the ancillary benefits of abatement policies in developing countries tend to suggest relatively high estimates which must be expected to reflect the relatively high local emission intensity of energy systems in many of these countries.

The study by Halsnaes and Markandya (1999) concluded that in some cases there might be employment generation benefits of some climate change mitig-ation projects that in the same way as local environmental side-impacts are significant compared with the abatement cost of the projects. The employment impact is, in the same way as local air pollution impacts, very site and context specific.

The application of a broader range of SD impact indicators to CDM projects as done in country studies for Brazil, China and India further emphasized the conclusion that there might be significant local side-impacts to the policies. A number of the projects identified suggest a lot of synergies between global and local policy priorities, but a number of the cases also show potential trade-offs. This is shown, for example, in the forestry sector analysis for Brazil. Low-cost plantation projects here are assessed to have negative impacts on the local environment, while other projects focus on native forest management generally have more positive local SD impacts, but are at the same time more costly, which can make them less attractive to donor countries.

On this basis, it can be concluded that empirical studies show that significant potential local environmental benefits and/or broader SD benefits connected to some climate change mitigation projects, but the magnitude of these are very project and context specific, and some projects can even have negative impacts on SD.

GENERAL CONCLUSIONS

Sustainable development is by definition a very broad conceptual framework that includes a wide range of short- and long-term policy goals that are highly relevant to climate change policy analysis. Many authors have introduced it as a general approach for assessing climate change mitigation studies, and recently this has introduced a general discussion about how to operationalize this conceptual framework.

The application of the SD concept to climate change studies has thrown up a number of theoretical and practical complexities. These are related to the very broad policy agenda that SD by definition is trying to address and to the very ambitious scope of the work. This scope is formulated by the IPCC TAR as enabling the assessment of the synergies and trade-offs involved in the pursuit of multiple goals – environmental conservation, social equity, economic growth, and poverty eradication (Banuri et al, 2001, pp103–104).

The current review has not identified any studies that are addressing how the broad agenda of SD policy objectives can be integrated comprehensively in climate change mitigation studies, but it has identified a number of studies that make a first attempt at integrating a number of important SD indicators. These studies include ancillary benefit studies and studies that have assessed climate change mitigation projects in relation to selected SD impact indicators representing development, environmental, social, and in some cases, technology implementation dimensions of sustainable development.

The ancillary benefit studies have focused on short-term side-impacts of climate change mitigation policies, in particular local air- and water-quality impacts. The SD impact studies have also included these local environmental impacts, but have suggested a number of additional indicators to reflect development such as employment generation, and social impacts, such as income distribution. The approach of ancillary benefit studies has typically been a cost-benefit or cost-effectiveness analysis, while the SD impact studies have been based on various approaches, including multicriteria analysis with quantitative and qualitative information, and cost-benefit analysis.

The review of empirical studies has identified many similarities in analytical approaches and suggested SD impact indicators across papers. Many studies have demonstrated that there is a potentially wide range of climate change mitigation policies which have significant co-benefits on the local environment and on development. These benefits can be very significant in a number of developing countries, especially those characterized by high local pollution levels and unemployed labour. Some climate change mitigation projects might also have a number of negative side-impacts on the local environment and employment. A strong conclusion across all the studies is that the impacts of climate change mitigation projects on SD are very site and context specific, and the actual impact from case to case should be assessed in relation to national development priorities.

A number of methodological issues have been highlighted as important to the establishment of more consistent information about SD impacts of climate change mitigation policies. These include an outline of the main linkages between SD and climate change mitigation, including development, environmental and social dimensions, in order to established consistent and operational SD impact indicators addressing intra- as well as intergenerational sustainability aspects. Another need is to develop a methodological framework that facilitates conceptual and analytical consistency across SD impact studies. The main components in such a framework would be the definition of cost concepts and SD impact indicators, baseline scenario approaches and assumptions, the terminology and format of qualitative and quantitative information, and the classification of analytical approaches.

NOTES

1   This literature has in many cases been inspired by the UN Framework Convention on Climate Change (UNFCCC), which in its article 3.4 emphasizes 'parties have a right to, and should promote sustainable development'.
2   Some approaches take the starting point in economic, ecological or social assessment approaches and include a broader set of policy variables that reflect broader SD impacts, while other approaches introduce new specific approaches for integrated assessment of SD impacts (Munasinghe, 2000).
3   IPCC TAR is not developing new literature but is assessing internationally peer reviewed literature. Conclusions presented in the IPCC TAR report is therefore representing an assessment of this literature and not new literature in this area.
4   The relative magnitude of the ancillary benefits of small and large GHG emission-reduction targets must be expected to vary significantly for various reasons. The marginal GHG abatement costs are expected to rise exponentially with increasing reduction targets, while the ancillary benefits of, for example, local pollution reductions must be expected to decrease with far-going reductions.
5   Dessus and O'Connor, 1999; O'Connor, 2000; Cifuentes et al, 2000; Garbaccio, Ho and Jorgenson, 2000; Wang and Smith, 1999; Aunan, Aaheim and Seip, 2000; Brendemoen and Vennemo, 1994; Barker and Rosendahl, 2000; Scherega and Leary, 1993; Boyd, Krutilla and Viscusi, 1995; Abt, 1999; and Burtraw et al, 1999.

REFERENCES

Abt Associates and Pechan-Avanti Group (1999) *Co-Control Benefits of Greenhouse Gas Control Policies,* prepared for the Office of Policy, US Environmental Protection Agency, Contract No 68-W4-0029
Aunan, K, Aaheim, H A and Seip, H M (2000) 'Reduced damage to health and environment from energy saving in Hungary', in OECD (2000) *Ancillary Benefits and Costs of Greenhouse*

Gas Mitigation, proceedings of an IPCC Co-sponsored workshop held on 27–29 March in Washington, DC

Austin, D, Faeth, P, Serôa da Motta, R, Ferraz, C, Young, C E F, Ji, Z, Junfeng, L, Pathak, M, Shrivastava and Sharma, S (2000) *How Much Sustainable Development Can We Expect From The Clean Development Mechanism?*, World Resources Institute, Washington, DC

Banuri, T, Weyant, J, Akumu, G, Najam, A, Pinguelli, R, Rayner, S, Sachs, W, Sharma, R, Yohe, G (2001) 'Scope of the report: Setting the stage: Climate change and sustainable development' in B Metz, B, 0 Davidson, R Swart and J Pan (2001) *Climate Change 2001: Mitigation Contribution of Working Group III to the Third Assessment Report of the Intergovernmental Panel on Climate Change*, Cambridge University Press, pp73–114

Banuri, T and Gupta, S (2000) 'The Clean Development Mechanism and sustainable development: An economic analysis' in P Gosh (ed) (2000) *Implementation of the Kyoto Protocol*, Asian Development Bank, Manila

Barker, T, and Rosendahl, K E (2000) 'Ancillary Benefits of GHG Mitigation in Europe: SO_2, NO_x and PM_{10} reductions from policies to meet Kyoto targets using the E3ME model and ExterneE valuations' in OECD (2000) *Ancillary Benefits and Costs of Greenhouse Gas Mitigation*, proceedings of an IPCC co-sponsored workshop held on 27–29 March in Washington, DC

Boyd, R, Krutilla, K and Viscusi, W K (1995) 'Energy Taxation as a Policy Instrument to Reduce CO_2 Emissions: A Net Benefit Analysis', *Journal of Environmental Economics and Management*, Vol 29, No 1, pp1–25

Brendemoen, A and Vennemo, H (1994) 'A Climate Treaty and the Norwegian Economy: A CGE Assessment', *The Energy Journal*, Vol 15, No 1, pp77–91

Burtraw, D, Krupnick, A, Palmer, K, Paul, A, Toman, M and Bloyd, C (1999) *Ancillary Benefits of Reduced Air Pollution in the US from Moderate Greenhouse Gas Mitigation Policies in the Electricity Sector*, Discussion Paper 99–51, Resources for the Future

Cifuentes, L, Sauma, E, Jorguera, H, Soto, F (2000) *Preliminary Estimation of the Potentisal Ancillary Benefits for Chile*, in OECD (2000) *Ancillary Benefits and Costs of Greenhouse Gas Mitigation*, proceedings of an IPCC co-sponsored workshop held on 27–29 March in Washington, DC

Dessus, S and O'Connor, D (1999) *Climate Policy without Tears: CGE-Based Ancillary Benefits Estimates for Chile*, OECD, Paris (draft)

ExterneE (1995) *ExterneE, Externalities of Energy*, Vols 1–7, European Commission, Directorate-General XII, Science, Research and Development

ExternE (1997) *Externalities of Fuel Cycles 'ExternE' Project: Results of National Implementation*, Draft Final Report, Commission of the European Communities, DGXII, Brussels

Garbaccio, R F, Ho, M S, Jorgenson, D W (2000) 'The Health Benefits of Controlling Carbon Emissions in China' in OECD (2000) *Ancillary Benefits and Costs of Greenhouse Gas Mitigation*, proceedings of an IPCC co-sponsored workshop held on 27–29 March in Washington, DC

Halsnaes, K, Callaway, J M and Meyer, H J (1998) 'Economics of Greenhouse Gas Limitations, Main Reports', *Methodological Guidelines*, UNEP Collaborating Centre on Energy and Environment, Risø National Laboratory, Denmark

Halsnaes, K and Markandya, A (1999) *Comparative Assessment of GHG Limitation Costs and Ancillary Benefits in Country Studies for DC's and EIT's*, paper presented at IPCC workshop on Costing Methodologies held on 29 June–1 July 1999 in Tokyo

Iyer, M (1999) *Removing the Myth of Sustainable Development in CDM: First Step Towards Operationalization*, IGES, Japan

James, B and Spalding-Fecher, R (1999) *Evaluation criteria for assessing climate change mitigation options in South Africa*, Energy & Development Research Centre, University of Cape Town (draft version 3)

Ji, Z and Jungfeng, L (2000) 'China: CDM opportunities and benefits', in Austin, D and Faeth, P (eds) *Financing Sustainable Development with the Clean Development Mechanism*, World Resources Institute, Washington, DC

Krupnick, A, Burtraw, D and Markandya, A (2000) 'The Ancillary Benefits and Costs of Climate Change Mitigation. A Conceptual Framework', in *Ancillary Benefits and Costs of Greenhouse Gas Mitigation,* proceedings of an IPCC co-sponsored workshop held on 27–29 March in Washington, DC

Markandya, A (1998) *The Indirect Costs and Benefits of Greenhouse Gas Limitiations,* Handbook Reports, The Economics of Greenhouse Gas Limitations, UNEP Collaborating Centre on Energy and Environment, Risø National Laboratory, Denmark

Markandya, A, Halsnaes, K (co-lead authors) and Lanza, A, Matsuoka, Y, Maya, S, Pan, J, Shogren, J, Serôa da Motta, R (Brazil), Zhang Tianzhu (China) (lead authors) (2000) 'Costing Methodologies', Chapter 7 in B Metz, O Davidson, R Swart and J Pan (2001) *Climate Change 2001: Mitigation. Contribution of Working Group III to the Third Assessment Report of the Intergovernmental Panel on Climate Change,* Cambridge University Press, pp451–498

Metz, B, Davidson, O, Swart, R and Pan, J (2001) *Climate Change 2001: Mitigation, Contribution of Working Group III to the Third Assessment Report of the Intergovernmental Panel on Climate Change,* Cambridge University Press, pp73–114

Munasinghe, M (2000) 'Development, Equity and Sustainability (DES)', in IPCC supporting material, Guidance Papers on the Cross Cutting Issues of the Third Assessment Report of the IPCC, IPCC, Geneva

O'Connor, D (2000) 'Ancillary Benefits Estimation in Developing Countries: A Comparative Assessment' in *Ancillary Benefits and Costs of Greenhouse Gas Mitigation,* proceedings of an IPCC co-sponsored workshop held on 27–29 March in Washington, DC

OECD (2000) *Ancillary Benefits and Costs of Greenhouse Gas Mitigation,* proceedings of an IPCC co-sponsored workshop held on 27–29 March in Washington, DC

Pathak, M, Srivastava, L and Sharma, S (1999) 'Indian Case', in Austin, D and Faeth, P, (eds) *How Much Sustainable Development Can We Expect from the Clean Development Mechanism?,* World Resources Institute, Washington, DC

Rowe, R, Bernow, S, Chestnut, L and Ray, D (1996) *New York State Externality Study,* Oceana Publications Inc., Dobbs Ferry, NY

Serôa da Motta, R, Young, C E F and Ferraz, C (2000) 'Brazil: CDM opportunities and benefits', in Austin, D and Faeth, P (eds) (2000) *Financing Sustainable Development with the Clean Development Mechanism,* World Resources Institute, Washington, DC

Scherega, J D and Leary, N A (1993) 'Costs and Side Benefits of Using Energy Taxes to Mitigate Global Climate Change, *Proceedings 1993, National Tax Journal,* pp133–138

Thayer, M A et al (1994) *The Air Quality Valuation Model,* Report to the California Energy Commission, Sacramento, CA, Regional Economic Research Inc. and TRC Environmental Consultants

Thorne, S and La Rovere, E L (1999) Criteria and Indicators for Appraising Clean Development Mechanism (CDM) Projects, Working Paper

UNFCCC (1997) *Kyoto Protocol to the United Nations Framework Convention on Climate Change (UNFCCC),* FCCC/CP/1997/L.7/Add.1, Bonn

Van Pelt, M J F (1993) Ecologically sustainable development and project appraisal in developing countries, *Ecological Economics,* Vol 7, No 1, pp19–42

Wang, X and Smith, K (1999) *Near-term Health Benefits of Greenhouse Gas Reductions: A Proposed Assessment Method and Application in Two Energy Sectors of China,* World Health Organization, Geneva

Chapter 4

Assessing Social Capital Aspects of Climate Change Projects

Anne Olhoff

INTRODUCTION

The social aspects of development and the concept of social capital are issues that currently receive a considerable amount of attention in various research disciplines and policy fora. As we saw in Chapter 2, including the discussion on social capital in the assessment of climate change mitigation policy is a task that is being attempted by a number of scholars. In this chapter, we look into the various ideas associated with social capital, and the main results of the literature on social capital in detail. Our objective is to analyse how specific aspects of social capital, seen from an economic perspective, may be included in the assessment of greenhouse gas (GHG) emission-reduction options.

Climate change mitigation policies have, as a major objective, the need to use existing resources more efficiently. One way of increasing efficiency is through technology substitution and the introduction of new technologies. In many cases, however, efficiency gains may also arise from, and depend on, structural changes and general improvement in the functioning of the economy, including the functioning of formal and informal markets. Since this source of inefficiency is a fundamental characteristic of many developing countries and since the primary objective of climate change mitigation strategies is to meet global climate change goals as efficiently as possible, the strategies have to be related to general development aspects and consequently to have the potential to assist in the realization of national development goals. The promotion of energy efficiency measures and renewable energy technologies may improve rural energy supply, energy efficiency in domestic industries and the reduction of air pollution. Similarly, climate change projects may have implications for other development priorities regarding, for example, employment, afforestation, land degradation, health, market development, information flows, the reduction of transaction costs, and education and learning effects. As will be elaborated in the following pages, the achievement of many of these goals depends on social capital.

Different factors hamper the implementation of climate change mitigation projects and the realization of project objectives. These are frequently described as barriers and provide one explanation why the recognized potential for climate change mitigation options with no or very low cost is rarely realized. We argue, however, that the general approach to barriers and barrier removal suffers from a weakness because it does not facilitate an assessment of how the barriers are interrelated and rooted in the social organization of society. The basic problem is that a coherent framework for assessing the overall functioning of the economy is missing. The inclusion of social capital aspects permits the development of such a framework and provides a link between climate change mitigation implementation issues and the concept of social capital. The chapter has two main objectives. One is to develop an analytical framework for the assessment of social capacity and apply the framework to the implications of social capacity for the implementation of GHG emission-reduction options. The other is to outline how cost concepts related to social capital and social capacity at the project level can be developed.

At the most general level, social capital is referred to as the 'glue' that holds society together (Narayan, 1999) and it is embodied in the relations between and among people (Coleman, 1990). While there is a lack of consensus regarding how social capital should be defined and measured (see below), a main contribution of the concept is to draw attention to the significance of the social organization of society for economic development.[1] In this way, the emphasis on social capital and more generally on the social aspects of development may be seen as a response to the limits of the traditional economic development paradigm in focusing on reforming the state and markets in order to achieve economic development. Such approaches to project assessment include cost-benefit analysis, bottom-up studies and sector studies. Social aspects are rarely taken into account in them.

Studies of social capital within economics analyse the ways in which social relationships affect economic outcomes and economic efficiency, and the externalities created by these relations.[2] The economics-related research within the field of social capital theory can be defined at two levels: the microeconomic and the macroeconomic level. At the microeconomic level, analysing the functioning of markets and institutions, as well as the links between social relationships and markets, are ways of approaching the issue of social capital. At the macroeconomic level, the focus is on how macroeconomic performance is affected by institutions, legal frameworks and the government's role in organizing production (World Bank, 1997b; Dasgupta and Serageldin, 1999). These aspects have important implications for the overall level of GHG emissions, as well as the intensity of GHG emissions per unit of activity.

Recent work attempts to combine the two approaches, focusing on the nature of relationships between and within groups and between the state and society.[3] These aspects are elaborated and a simple analytical framework for the assessment of social capacity is developed. The analysis is supported by insights from economic theory. There is a strong correspondence between the parts of the social capital theory being studied and results from economics addressing the same issues. In the later parts of the chapter, the framework is applied specifically to the analysis of the social capacity base for implementation of GHG emission-reduction projects. The implications for specific social capacity-building efforts and the associated costs at the project level are then explored. Finally, some general conclusions are provided.

Even though the social capital aspects are addressed from a project implementation perspective, they are still relevant for host country selection of projects in that the basic issue being addressed is development. Understanding and enhancing the social capacity of society provides a general potential for reducing transaction costs (through, for instance, education, learning effects, the spreading of information, and technology transfers), private sector and market development, and efficiency gains that reaches beyond the project level.

GHG EMISSION REDUCTION OPTIONS AND THEIR IMPLEMENTATION

Large-scale GHG emission-reduction requires energy-efficient resource utilization and renewable energy consumption. One way of increasing efficiency is through technology substitution and the introduction of new technologies.[4] In many cases, however, efficiency gains may also arise from and depend on structural changes and general improvement of the functioning of the economy. These aspects are directly related to more broad development aspects. In relation to this, it is often noted that GHG emission options in many cases represent win-win opportunities because they result in both global and domestic benefits.[5] In the case of energy efficiency, global GHG emissions are reduced and domestic resource utilization is improved simultaneously. Options to increase energy efficiency will in several cases also be directly related to domestic development priorities because they influence factors such as local indoor and outdoor air pollution directly related to health objectives. At a more general level, the options may create synergies between development priorities and environmental concerns because they have the potential to address market failures – thereby increasing efficiency – and improve the environment at the same time.

The implementation of different options will generally involve the private sector as well as be based on the interaction between the private and public sectors of the economy. In developing countries, the (formal) private sector is often weak and segmented. From an implementation perspective, this suggests that it may be appropriate to involve governments and informal institutions, at least in the initial phases of implementation, as well as developing the private sector.

More specifically, markets are important in relation to GHG emission-reduction options because their implementation as a general rule necessitates the assistance of formal and/or informal markets. This can be illustrated by looking at the examples of mitigation options in different sectors given in Table 4.1.

The importance of markets can be illustrated by studying some specific cases. Options related to end-use efficiency in different sectors include the replacement of incandescent bulbs with efficient fluorescent and compact bulbs in the residential sector, the use of improved cook-stoves and efficiency improvements for vehicles. The implementation of such options based on a traditional economic assessment presupposes not only that markets exist but also that they are fairly well functioning.

'Well functioning' in this case requires that:

- Information about technology characteristics, energy-savings potential, economic and financial costs and benefits, and so on is transmitted to all relevant agents
- Credit is available
- Uncertainty about future energy prices and government policies is not too high
- Consumer discount rates are not unduly higher than prevailing market discount rates or than discount rates for energy supply investments
- Transaction costs related to identification, procurement, operation, maintenance and installation do not prohibit undertaking the transactions

These aspects are related in turn to the performance of institutions other than markets. Information will be transmitted through different channels such as television, newspapers, work-related networks, informal groups and civil society groups, and may also be provided by specific information campaigns. Access to credit will depend on the nature of formal and informal institutional arrangements. Similarly, uncertainty about government policies is related to the overall accountability of the formal institutions. Transaction costs and consumer discount rates will depend on the just-mentioned institutional factors, as well as, for example, the existing educational systems in the case of transaction costs and the general income level in the case of discount rates.

The achievement of GHG emission reduction through more efficient resource utilization and structural changes may thus be constrained by weak markets and institutional structures because these factors reflect the general social capacity for changes in a given situation. Many studies of the opportunities for different technology options, including technology transfer, are based on experience from industrialized countries where social capacity to a large extent supports policy implementation. It seems reasonable to assume that the factors related to social capacity will be more binding in a developing country context than they are in industrialized countries. As will be discussed in more detail in the later parts of the chapter, a generally weak institutional structure and the associated presence of high transaction costs may have important implications for trade, specialization, efficiency and consequently the development of markets. If markets are thin or missing, trade is on a very small scale and specialization in areas of comparative advantage does not take place, growth is hindered and allocative efficiency is not reached.

Under these circumstances, mainstream economics does not apply very well, as the assumptions necessary for most economic results and models are violated. Thus, dysfunctional institutions can be seen as one of the reasons why mainstream economics does not apply as readily in developing countries as it does in developed countries.

The above also suggests that the successful implementation of GHG emission-reduction options in most cases will depend on additional measures to increase the potential market and the number of exchanges. This can involve strengthening the incentives for exchange (prices, capital markets, information efforts and the like), introducing new actors (institutional and human capacity efforts) and reducing the risks of participation (legal framework and information and the

Table 4.1 *Examples of mitigation options in different sectors*

Energy sector
  End-use efficiency improvements in household, industry, service.
  Transmission systems
  Fuel substitution
  Renewable technologies (decentralized)
  Supply technologies (centralized): fossil fuels, nuclear and renewable.
Agricultural sector
  Fertilizer control schemes
  Introduction of crops with enlarged carbon sequestration capability
  Livestock management: manure treatment, feeding
  Cultivation of rice paddies
Forestry sector
  Afforestation projects
  Increasing the carbon sequestration capability of growing forests (increasing biomass density)
  Recycling or permanent storage of carbon sequestered in harvested biomass
  Reforestation
Transportation
  Efficiency improvements for vehicles
  Switch to fuel systems with lower emissions
  Improve transport system efficiency
  Modal shifts
  Manage transport demand
Waste management
  Gas recovery from landfills
  Biogas plants
  Recycling
  Composting
  Industry
  Cement production
  Aluminium production

Source: Halsnaes, Callaway and Meyer (1998)

general policy context of market regulation). The measures all depend on the nature of the formal institutions, the social groups of society and the interaction between them. These factors are in turn important determinants of the social capacity for changes in a given context. Implementation policies should therefore reflect the prevailing social capacity and might entail social capacity-building.

By social capacity-building we mean activities that enhance the capacity of individuals, institutions and society to undertake an activity or project that is in accordance with general economic policies and development programmes. In this way, the capacity-building efforts become catalysts for the market development and commercialization of the specific options under consideration. As a consequence, social capacity-building efforts are assumed to be temporary activities that establish a permanent capacity for a given option.

To summarize, by addressing overall efficiency, GHG emission reduction touches a fundamental aspect of development. The implementation of different GHG emission-reduction options is closely related to the social capacity for changes, which in turn is a function of factors such as formal and informal institutions – including but not confined to markets – and the relations between these institutions.

Consequently, the central theme of the chapter is the interlinkage between social capital, markets and the implementation of GHG emission-reduction options. More specifically, the following issues will be discussed:

- The extent to which informal institutions can make up for deficiencies in markets
- The evolution of social capital – does it relate to development stage and does its type change as society transforms?
- Links between macrolevel institutions and microlevel institutions. When do they reinforce each other and when do they displace each other?
- The relation between markets and other institutions. Do formal and informal institutions tend to prevent markets from developing or functioning?
- The conditions under which the institutional foundation supports the implementation of a project or strategy and situations where additional measures are required for successful implementation

SOCIAL CAPITAL, INSTITUTIONS AND THE NOTION OF LINKAGES

A frequently cited definition of social capital is that of Putnam, Leornardi and Nanetti (1993, p67) as those 'features of social organization, such as trust, norms, and networks that can improve the efficiency of society by facilitating coordinated actions'. More generally, a common denominator of the definitions of social capital is that they emphasize features embodied in institutions, where institutions are broadly defined as fora with implications for rules of interaction, and include family, kin, networks and markets.

In the present context, we concentrate on two aspects of social capital, the notion of linkages between and within groups, and the interaction between state and society.[6] The argument is that it is the nature and extent of institutions and the interaction between them that determines the social capacity of society to implement, among other things, projects such as GHG emission-reduction projects. The social capacity to implement projects is, in other words, reflected in the composition of social capital in a given society at a given time. One of the main aspects of interest in this context is the ways in which formal markets are influenced by and in turn influence the social capital aspects.

By focusing our attention on the institutional aspects of social capital, we avoid the problem posed by most definitions of social capital of amalgamating objects such as trust (beliefs), norms (behavioural rules) and networks (capital assets) to which there is no simple solution (Dasgupta,1999).

While this approach to social capital is narrow, we argue that it captures the central aspects for the development of a simple analytical framework for the assessment of the social capacity to implement GHG emission-reduction projects. The discussion of the social capital aspects are supplemented and supported by findings from economic literature addressing the same issues. The key elements of the framework are provided in the last part of this section.

Linkages in Social Capital Theory

Social groups are found at all levels in all societies and are the central unit of analysis in social capital theory. The basic idea of social capital is that these groups constitute an important asset, not only at the individual level but also at the community and society level. The social groups, networks, associations – or, for short, informal institutions – facilitate collective action by increasing information flows and reducing transaction costs, thereby enabling the achievement of outcomes that would not be possible in their absence.

Development and Linkages at the Microlevel

Social groups may be based on class, caste, ethnicity, family, geographical location, work, mutual interests, and so on, and will differ in their access to power and resources. In the following, the form of social capital that brings people who already know each other closer together, will be referred to as bonding social capital, represented by intragroup linkages. Membership of a group is a source of protection and risk management, especially for the poor, and serves basic solidarity functions (Kozel and Parker, 2000).

Apart from the intragroup linkages in the form of families, friends and church groups, examples of bonding social capital include community groups managing common property resources, small-scale credit schemes and farmer groups. Bonding social capital thus influences practices and resource-use patterns that may be of relevance in relation to climate change options. A high density of social groups – ie, a large stock of bonding social capital is, however, not necessarily desirable. Intragroup linkages are associated with costs as well as with benefits.[7] The exclusion of social groups places high costs on the individual. Bonding social capital may also preserve inequality in society by constraining economic and social mobility, and the non-economic claims such as obligations and loyalty of this type of social capital may impose negative economic consequences by preventing the individuals from breaking free and/or joining other groups. Finally, a high degree of bonding social capital keeps transactions at a minimal level. In this sense, bonding social capital may act as a 'straitjacket' to development.

Furthermore, several authors have stressed that there are other possible costs to consider. Social groups may work against the overall interests of society, as is the case with the Mafia, drug cartels, street gangs, and the like.[8] Likewise, nepotism may be associated with the presence of strong social groups. This points to the importance of governance and formal institutions, to which we will return below.

Whether the costs or benefits of bonding capital prevails in a given situation at a given time, may be assessed by considering the presence or absence of a complementary form of social capital that bring together people or groups that previously did not know each other. We call this form of social capital bridging social capital and it is characterized by cross-group linkages.[9] These linkages contribute to the benefit side because by cutting across social groups and between social groups and government, they open up economic opportunities to those belonging to less powerful or excluded groups, and provide a source of economic and social mobility. Similarly, the linkages facilitate transactions between groups and thus the expansion of the scale of economic activities.

Examples of bridging social capital include civil society and community groups built to provide credit opportunities, basic healthcare and education services at the community level, risk management, and to manage common property resources, as well as solving local disputes. Entrepreneurial networks are also examples of cross-group linkages. By providing linkages between groups, bridging social capital facilitates the transmission of information related to, for instance, markets, products and technologies, prices, and state performance. In relation to climate change mitigation, the implementation of options such as new technologies for increasing end-use efficiency may be facilitated by using the information-sharing role of the cross-group linkages.

Several authors have pointed out that bottom-up development – ie, development initiated at the local (or micro) level, is a function of both intragroup linkages and cross-group linkages, and that the presence of both forms of social capital are necessary conditions for economic development. As early as 1908, Georg Simmel (1971) notes that poor communities need to generate linkages beyond their group to achieve long-term developmental outcomes.

In the same line of thinking, Granovetter (in Portes, 1995) asserts that economic development takes place through a mechanism described as coupling and decoupling. The mechanism – that for instance has been used to describe the occurrence of Rotating Savings and Credit Associations (RoSCAs) – allows individuals to draw initially on the benefits of intragroup linkages (coupling), while also enabling them to acquire the skills and resources to participate in networks that transcend their group or community – ie, create cross-group linkages (decoupling). Put differently, the stock of social capital in the form of intragroup linkages can serve as a basis for initiating development, but it must be supplemented over time by other forms of capital, notably cross-group linkages. The latter is essential for the social and economic mobility required for economic development. One example of this is that social impediments to mobility in general and particularly labour mobility can hinder technological progress, as talents are not able to find their ideal locations which results in inefficiency (Dasgupta, 1999). Several studies also find that a high density of bonding social capital does not necessarily have positive effects on poverty. In Kenya, for instance, a participatory poverty assessment recording more than 200,000 community groups in rural areas, concluded that the effects of these groups on poverty were negligible, one reason being the lack of cross-group linkages (Narayan and Nyamwaya, 1996). Similarly, while indigenous groups in Latin America are often characterized by a high degree of bonding social capital, the lack of cross-group linkages may explain why these groups remain excluded economically (Narayan, 1999).

Building on Granovetter, Woolcock (1998, p175) identifies three main reasons why business groups in poor communities need to generate and maintain cross-group linkages:

- They enable the group to resist the economic and non-economic claims of community members when they undermine (or threaten to undermine) the group's economic viability and expansion or put differently, in situations where the costs of the intragroup linkages are becoming larger than the benefits
- They can secure entry to more sophisticated factor and product markets
- They enable individuals of superior ability and ambition within the business group itself to enter networks that are larger and more complex than the network based on intragroup linkages

In a study of enterprise performance in Ghana, Barr (1998) brings evidence to support the above. She shows that while networks of small entrepreneurs[10] relying on bonding social capital (what she calls solidarity networks) reduce information asymmetries and support informal credit and risk-sharing arrangements, thereby reducing risk and uncertainty, they have little overall effect on enterprise performance, as measured by income-earning capacity and competitiveness. It is the so-called innovation networks of larger enterprises, built to increase productivity, profits and market share, and relying on bridging social capital to provide information about technology and markets that have significant effects on enterprise performance. The main explanation for this is that in general the networks of the poor are defensive in nature, in that their main objective is survival, while the networks of the non-poor are aggressive in nature – ie, they aim at expansion.

The study shows a marked difference in the procurement of formal credit of the two types of enterprises. Where less than 20 per cent of the small enterprises have a formal loan or overdraft, the percentage is 75 of the larger enterprises. These figures reflect the imperfect nature of the formal credit and insurance institutions, as well as the role of networks as substitutes for formal market-supporting institutions – in this case the risk and uncertainty reducing function of the solidarity networks referred to above.[11]

Furthermore, it directs our attention to an additional feature associated with a gradual shift in the significance of intragroup linkages and cross-group linkages in the course of development. Earlier we noticed that the costs and benefits associated with a given combination of bonding and bridging social capital are context and time specific, in that they depend on the developmental stage of society as a whole. As long as communities remain small and transactions take place among known individuals with repeat deals common, the costs of transactions are minimized. The problem is that intragroup linkages also minimize the gains from trade. The development of cross-group linkages facilitates an increase in the volume of transactions but is at the same time associated with a change in the basis of transactions. Transactions based on intragroup linkages typically operate on norms of reciprocity (see, for instance, Dasgupta and Serageldin (1999) or Kranton (1996)), whereas cross-group linkages are less cohesive and more impersonal. As a result, it becomes more difficult to impose sanctions in the case of deviant behaviour. This relates well to the distinction made by Putnam, Leornardi and Nanetti (1993) between horizontal and vertical networks, in that the latter, according to the authors, cannot sustain trust and cooperation. To realize the full potential of cross-group linkages, their development should accordingly be assisted by a gradual shift in the relative importance of informal and formal institutions. Formal institutions provide the basis in the form of laws, rules and regulations for enforcing the less personal transactions based on cross-group linkages.

Development and Linkages between Macro- and Microlevel

Uphoff (1992, p272) points out that top-down efforts – ie, efforts initiated at the macrolevel – usually are needed to introduce, sustain and institutionalize bottom-up development. Both are necessary and positively related to the achievement of development.

It is at the state level that the enabling environment for social capital formation at the microlevel is shaped by factors such as policies; laws, rules and regulations; the provision of infrastructure, healthcare, educational systems; and the transmission of information. Finally, the capacity and credibility of the formal institutions in administering and enforcing these factors, as well as the ability to generate links between groups of citizens and public officials, affects the enabling environment. In turn, the informal institutions will influence government policies and government performance.

Among the authors of the macrolevel approach to development is Douglass North. According to North (1990), it is the institutional structure at the macrolevel that determines the movement from informal exchange based on networks to impersonal markets. The formal institutional setting determines the expansion of markets that leads to the division of labour and specialization required for enhanced economic performance.

The emerging synergy approach within social capital theory seeks to integrate micro- and macrolevel approaches to development by studying mutually supportive relations between governments and groups of citizens. Evans (1996) makes a distinction between two forms of synergy – complementarity and embeddedness – that are usually present at the same time to a higher or lesser degree.

Complementarity represents the conventional way of conceptualizing state–society relations. The state provides the legal and political framework and is responsible for delivering certain public goods and services (cf the second paragraph of this section), which complements goods that are produced more efficiently by the private sector (or private actors).

In relation to climate change mitigation, a relevant example of the concept of complementarity is power sector reform. The overall objective of such reform is to improve economic and technical efficiencies within the power sector, thereby reducing GHG emissions, and the principal tools for realizing this objective have been liberalization and privatization. Typically, power sector reforms have involved reform of the regulatory systems (ie, liberalization) and/or changes in the structure and ownership of the sector (ie, privatization). These interventions influence various aspects of performance:

- Financial performance: cost-recovery and future investments
- Supply-side efficiency: the efficiency with which electricity/energy is produced and delivered to consumers

- Demand-side efficiency: the efficiency with which electricity/energy is used by the consumers

The concept of complementarity can be seen as part of the underlying rationale of reforms, as it reflects the principle that the state should provide the overall regulatory and legal framework, while the production is handled more efficiently by the private sector.

Embeddedness describes the ties that connect citizens and public officials across the public–private divide. This is directly relevant to the practical aspects of policy implementation because it concerns the ways in which the two types of agents interact at different levels of the implementation process. Often quoted is the example of irrigation, where local irrigation officials are integrated in the local social relations – and hence under pressure to perform – because they are from the community being served by the irrigation. This is a more novel idea as the networks are not perceived as instruments of rent-seeking or corruption. It requires, however, that the overall institutional environment be accountable and competent. As Evans (1996, p1125) concludes himself, the most obvious candidate for a 'missing ingredient' to synergy in developing countries is a competent and engaged set of public institutions, rather than the existence or emergence of social capital at the community level.[12] This can help to explain the emphasis in the literature so far on the negative aspects of such ties.[13]

Prominent examples of the lack of embeddedness include descriptions of rent-seeking, inefficiency and ineffectiveness in Soviet Russia, Eastern Europe and China, where the problem is described as too much bureaucracy and too little society (Woolcock, 1998). Similarly, among democratic countries, India is mentioned as an example of lack of embeddedness. Despite well-educated civil servants, the ties to society – for example, in the form of industry groups – are limited, particularistic and poorly coordinated. Lack of embeddedness may also be reflected in slow or no response to citizen demands, indifference to the conditions of vulnerable groups and the misallocation of resources.

Conversely, Japan, Singapore and South Korea provide examples of embeddedness, especially between government ministries and large business groups through ties which bind the state to society and provide institutionalized links for ongoing negotiations on goals and policies (Evans, 1992, p164; Evans, 1995).

Social Capital and Markets

Adding the state–society perspective to the discussion on bridging and bonding social capital permits some further comments regarding economic development. Previously, we noted that one aspect of development is a gradual expansion of the volume of transactions in society and that a shift in the importance of cross-group linkages relative to intragroup linkages facilitates such an expansion. As this process implies that transactions become less personal, we further noted that it is assisted by the presence of accountable, formal institutions.

A process such as the one just described fits well with the results in economics that growth stems from the gains of trade (Ensminger, 1992)). Cross-group linkages are, however, just one step in the process. The scale can be further increased. The growth of impersonal markets seems to be necessary for increasing the scale of transactions and for long-run improvements in the standard of living. Dasgupta (1999, p390) notes that both theory and empirics testify to this, since transactions limited to a group are likely to be less productive than are those that involve an entire population. In addition, it is possible to show theoretically that the more dissimilar are transactors, the greater are the potential gains from transaction. A similar result is found within the field of risk-sharing and informal credit. *Ceteris paribus*, these arrangements are more efficient the larger and more diverse the group (Alderman and Paxson, 1994).

Transaction Costs

There is, however, nothing automatic about such a process as transactions are associated with costs. In mainstream economics, the emergence of institutions is seen as a response to high transaction costs, and the effect of institutions is increased economic efficiency. This causality[14] is rejected by new institutional economists who stress that in developing countries, institutions change and emerge as a result of personal gains, rather than being based on what represents the most efficient outcome for the economy as a whole. This perception is supported by indications of a relatively weak formal institutional structure in many developing countries and in that the most obvious candidate for a missing ingredient to the success of the mainstream view in developing countries, as mentioned above, is a competent and engaged set of public institutions.

In relation to climate change, an example of the two types of causality mentioned above may be as follows. Suppose that there is a policy commitment to limit GHG emissions from the industrial sector to a given level. Initially this level of GHG emissions will be associated with high transaction costs, as no mechanism(s) for allocating GHG emissions exists. According to the first type of causality, the state will respond to the need for new institutions, as witnessed by high transaction costs. An economically efficient response in this case is, for example, to establish markets for pollution permits. But if the second type of causality prevails, the response may come, for instance, from influential segments of or lobbying groups in the industry sector. In this case, the allocation of GHG emissions takes places according to the power structure within the industry sector and the trading of emissions is rare. Consequently, the resulting allocation will only result in efficiency by accident .

Given the absence or unreliability of formal institutions, transaction costs are minimized by limiting the scale of exchange. This implies that there will be a tendency not to develop larger, impersonal markets because transaction costs prevent the process. An example may be food and woodfuel markets in rural areas of developing countries. The market exchange process, is among other factors, limited by weak infrastructure and exchange is subsequently geographically limited. Exchange will have a tendency to be confined to a limited number of small societies, such as local villages, families or ethnic groups, keeping transaction costs low. The problem is that the market and production scales are low, implying high production costs and weak demand.

One way of mitigating a poorly developed institutional environment and resulting weak market development is by vertical integration. Ensminger (1992) reports findings from studies in West Africa, where institutions meant to facilitate the flow of information and resolve disputes concerning credit, property rights, and so on, were unreliable or non-existent:

> *As a consequence. . . one found a high degree of vertical integration, with one organisation (an ethnic group) involved in many different stages of production and distribution of a particular commodity. Because transaction costs were otherwise high, members ensured a greater degree of trust and cooperation by establishing an ethnic trading diaspora, thus reducing costs over what they would have been if members had been forced to interact in the marketplace for all production and marketing components.* (Ensminger 1992, p26)[15]

While networks characterized by vertical integration, as, for example, long-distance trading in earlier times, can be a means by which markets get established (Dasgupta (1999, p87)), they can also distort competition. Facing vertical integration, a small trader will find it very difficult to be competitive in any segment of the trade. A weak institutional structure may therefore facilitate monopsony situations.

The above example of woodfuel markets in developing countries is relevant to the discussion of monopsony situations. While not explicitly addressing this issue, a study of the charcoal cycle in Ghana (Nketiah, Hagan and Addo, 1988) provides information about the procurement of wood for charcoal production, which indicates that this link in the charcoal cycle may be characterized by a monopsony situation and that the charcoal production in general may be characterized by vertical integration. The majority of the charcoal producers belong to two ethnic groups operating in different regions of the country. In most cases, charcoal producers get access to the wood resource by paying a small fee to the chief of a given area (or community), or alternatively to the individual farm owner. Given the lack of infrastructure for transportation and remoteness to production sites and markets, a (small) producer of wood in a given region of the country will in practice only face one buyer.

The above illustrates a point made earlier regarding the ways in which informal networks substitute for formal market-supporting institutions. All societies rely on a combination of impersonal markets and informal institutions that shifts through changing circumstances. In the absence of accountable, formal institutions, the ways that people or groups of people find of circumventing difficulties in realizing mutually beneficial transactions are primarily based on informal institutions. The studies of Barr (1998, 2000) among other issues point to the network's role as substitutes for imperfect formal credit and insurance institutions. In the same vein, Dasgupta (1999) notes that up to 90 per cent of the value of all loans in Nigeria were obtained from the informal sector as recently as in the late 1980s.

Externalities

A closely related aspect of the links between formal markets and informal institutions is that they are associated with externalities. An externality is present when transactions in one institution have effects that spill over to another without being accounted for. In this case, private costs and benefits differ from social costs and benefits. The following cases are considered here. First, the case where externalities lead to situations in which informal institutions and markets displace each other inefficiently is considered. Then the case of externalities arising from the displacement of informal arrangements is considered.

New institutional economists accept that people may well resist trade as long as the costs of exchange are higher than the perceived benefits. Kranton (1996) explores the case where some agents engage in exchange based on bonding and bridging social capital – for short, defined as informal exchange in the following – and the remainder in monetary market exchange. Engagement in market transactions requires search for trading partners, which involves costs. The search costs undertaken by one agent lowers the search costs facing other agents, thereby creating an externality. The agents engaged in the first type of exchange minimize their search costs but only obtain the good produced by their trading partner. The agents engaged in market transactions have higher search costs but at the same time access to a variety of goods. The market search costs depend on the rate of occurrence of informal exchange. The higher the rate of occurrence of informal exchange, the thinner the market and the higher the search costs. If, conversely, the market is thick and the search costs are low, it becomes difficult to enforce an informal exchange agreement, because the sanctions for reneging on such an agreement are not severe as the agent has the opportunity to engage in market exchange.

Owing to the negative external effects that both types of exchange have on the other, it is socially efficient to engage in either informal or market exchange, depending on certain external factors. When goods are more (less) substitutable, informal (market) exchange is socially efficient (Kranton, 1996, p831). It is shown, however, that the economy will not necessarily converge to the efficient mode of exchange. The outcome depends on the initial market size. In other words, if market exchange is widespread initially, informal exchange cannot be enforced and market exchange survives. If informal exchange is large initially, the market does not function well, which results in more agents forming informal exchange agreements and the survival of this mode of exchange, regardless of whether this type of exchange is the more efficient.

What are the implications of the analysis? First, applied to developing countries characterized by a relatively weak formal institutional structure, it is reasonable to assume that generally the initial market size will be low. In this case, informal institutions can prevent markets from functioning well and may even prevent the emergence of markets. In such cases, exchange-based informal institution are an impediment to economic development. This also illustrates that the effectiveness and efficiency of institutions depends on the circumstances and may change over the course of time. Second, the analysis suggests that informal exchange is more likely to persist when based on cross-group linkages than when based on intragroup linkages. As discussed previously, cross-group linkages provide the means for expanding the volume of transactions and can increase the variety of goods available, thereby making the informal exchange more competitive with formal markets. Third, when informal exchange arrangements are based on the production and exchange of marketable goods and exchange is frequent and repeated between the same agents, they can be destructively competitive with markets. Conversely, market exchange is more likely to persist in the case where the goods in question are heterogeneous and agents do not interact frequently because the sanctions for reneging on agreements that are available to informal exchange arrangements are missing under these conditions. Fourth, while informal exchange arrangements can reduce transaction costs in the short run, they may sustain or increase these costs in the longer run by posing obstacles to the emergence or evolution of institutions and organizations facilitating market transactions.

A study of informal institutional arrangements in credit, land markets and infrastructure in Trinidad and Tobago provides findings in line with the above (Pamuk, 2000). It is shown that informal institutions based on cross-group linkages (for example, rotating savings and credit institutions known as the sou-sou) support transactions by reducing transaction costs, lowering risk and providing the means to cope with uncertainty and that they are used instead of formal institutions. The study points out that policy-makers generally should expect this to occur.

From a policy perspective another inference can be made regarding the role of the state in influencing the links between formal markets and informal institutions. As suggested previously, the formal institutional structure provides the basis in the form of laws, rules and regulations for enforcing less personal transactions based on bridging social capital and impersonal market transactions. It influences the general functioning of markets and the search costs facing the individual agent. Policies aimed at reducing transaction costs, lowering risk, providing mechanisms for coping with uncertainty and increasing information flows may therefore change the results of the analysis presented above. Furthermore, such policies may be able to take advantage of the informal institutional arrangements. When networks are used for transmitting information, both among network members and between members and non-members, they can contribute to the good functioning of markets. In such cases networks and markets are complementary in their roles.

We now turn to a different aspect of externalities, which is addressed from a policy perspective. When markets displace informal institutional arrangements at the community level in the production of goods and services, there are people who suffer unless countermeasures are undertaken. The study of interlinkages between markets and informal institutions is one way of approaching this aspect.

Country studies for developing countries suggest that interlinkages from one transaction to another are common and may be an important factor in explaining policy failure. It is appropriate to talk of an interlinkage when the terms and conditions of one contract[16] affects the terms and conditions of another contract. A study of 110 villages in West Bengal suggests that the credit–labour interlinkage is important (Bardhan, 1981, 1985). The peasants get credit in the form of food from the landlord during one season and pay back, providing the landlord with labour in the busy season. This example has two implications worth emphasizing in the present context. First, interlinkages raise the entry barriers, thereby possibly increasing the monopoly power of the stronger party (in this case the landlord). The freedom and mobility of the peasants are constrained because of the interlinkage. As was mentioned earlier, this leads to allocative inefficiency with negative implications for economic development. Second, if policy-makers do not take the interlinkages into account, there is a risk that policies will fail. In this specific case, land reform programmes affecting labour market conditions may not succeed if they are not combined with credit reforms. If credit reforms are missing, the former tenant may not have access to credit and is simultaneously cut off from the previous contract with the landlord.

In relation to climate change mitigation policies, the example indicates that it may be appropriate to pay attention to the functioning of credit markets. A mitigation policy aiming at raising efficiency in the agricultural sector by the establishment of well-defined property rights, thereby reducing the pressure on marginal areas as well as forests, is more likely to succeed if it is combined with appropriate credit schemes. Similarly, improved access to credit may reduce deforestation by making other options available to the private agents.

The discussion thus suggests that the presence of high transaction costs and externalities may have important implications for trade, specialization, efficiency and consequently for the development of markets. If markets are thin or missing, trade is on a very small scale and specialization in areas of comparative advantage does not take place, growth is hindered and allocative efficiency is not reached.

Under these circumstances, mainstream economics does not apply very well, as the assumptions necessary for most economic results and models are violated. Thus, dysfunctional institutions can be seen as one of the reasons why mainstream economics does not apply as readily in developing countries as it does in developed countries.

The Analytical Framework

One inference of the synergy approach outlined above, which seeks to integrate micro and macro linkages by promoting mutually supportive relations between governments and groups of citizens, is that several developmental outcomes are possible. These outcomes will depend on the interaction between formal and informal institutions at different levels of society and the presence or absence of certain elements at respectively the micro- and macrolevel. To synthesize, social capital in the form of intergroup linkages and cross-group linkages represent such elements at the microlevel. At the macrolevel, the state–society relations, reflected in the government ability to provide, nurture and enforce an enabling environment for development is a crucial element.

Woolcock (1998) presents a range of such possible development outcomes. Narayan (1999) integrates the central ideas of synergy, focusing on social capital in the form of cross-group linkages and the functioning of the state characterized by the state–society relations. The notions of complementarity and substitution are introduced to characterize the interaction between state and society.[17] Complementarity describes situations where the formal institutions of the state and the informal institutions reinforce each other. If the state is dysfunctional, the informal institutions become a substitute for the formal institutions and are reduced to serving a defensive or survival function. Furthermore, in the relative absence of cross-group linkages, social cohesion as well as mobility is low, which constrains society in the pursuit of well-being.

The resulting framework in Narayan (1999) has a fourfold classification (ie, complementary and subsitution on one dimension, and the prevalence of bridging versus bonding capital on the other) of development states that may be applied to countries, regions or communities, and indicates that different interventions are needed for different combinations of bridging social capital and governance. The paper has the description of the overall political, social and economic characteristics of society at its centre and provides an overview of findings from the recent empirical literature to support the division suggested by the framework.

For the analytical framework of this chapter, we will adopt the distinction between complementarity and substitution and a prevalence of respectively bonding and bridging social capital.[18] Rather than focusing on overall political, social and economic performance, however, the combinations of these two dimensions of social capital are seen as reflecting the social capacity, resulting in different enabling environments for activities in a given country, region or community (see Figure 4.1 below). The term social capacity base is used to describe the state of the various enabling conditions. More specifically, the purpose of the framework is to analyse the implications of social capacity for implementation of GHG emission reduction projects.

The basic framework is illustrated in Figure 4.1. The combinations of the two dimensions of social capital – ie, the state–society relations and the degree to which bridging social capital is present – result in differences in the social capacity. Drawing on the previous parts of the section, the main characteristics of the enabling environments of the resulting quadrants are respectively labelled optimal conditions (note that they are not strictly optimal in the economic sense of the word), microlevel challenges, minimal conditions and macrolevel challenges.

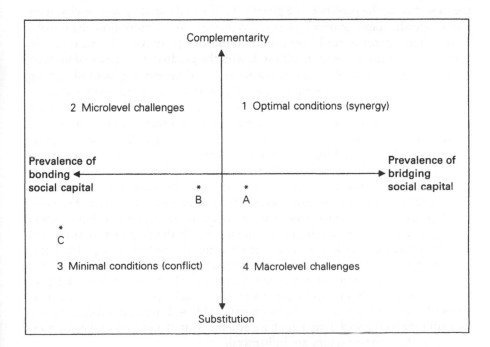

Figure 4.1 *Social capacity and enabling environments*

The different enabling environments indicated by the analytical framework reflect variations in the social capacity for the implementation of policy options at a given level of society. The framework is applied to the analysis of GHG emission-reduction options later in this chapter, where each of the quadrants will be elaborated on in turn, synthesizing the discussion of this section. However, a few general comments can be made at this point.

As Figure 4.1 indicates, each of the enabling environments of the quadrants is characterized by a continuum of possible combinations of the social capital aspects under consideration. The variations in enabling conditions can be at least as significant, and may in some cases be more significant, within a given quadrant as they are between quadrants. This can be illustrated by considering a move from point A to point B in the figure. While this implies a shift in the overall category of an enabling environment, the difference in social capacity between the two points is not dramatic and arises from a relatively small change in the prevalence of bridging social capital. While remaining within the same overall category of an enabling environment, a move from point B to a point such as C has more significant implications for social capacity, because the changes in the functioning of the state as well as in the nature of the prevailing informal groups are much more profound.

While this should be kept in mind in the discussion of the quadrants below, the policy options and recommendations will depend in general on which of the four categories of social capacity that best describe a given situation.

Another aspect related to the above is that social capacity is a dynamic concept. Throughout the section, we have emphasized that the different aspects of social capital and their interrelations change in the process of development. By the same token, the presence of a given type of social capacity will be characteristic of specific stages of development. A crude textbook description of a possible evolutionary process could be: initial conditions primarily characterized by bonding social capital as in quadrant 3, with the gradual emergence of bridging social capital, leading to a shift to quadrant 2, and the development of a strong formal institutional structure ultimately facilitating a move to quadrant 1.

Furthermore, general policies and specific options should not only be compatible with the social capacity characterizing a given situation. They should seek to improve on the enabling environments represented by the quadrants of the framework, thereby making a gradual transformation towards quadrant 1 possible. This also points to the possible developmental role of specific options, such as climate change mitigation options emphasized at the beginning of the chapter. The implementation of these options should not be in conflict with the overall development strategy. Furthermore, the specific options should exploit potential synergies with general development policies. While development policies with specific objectives such as redistribution generally will be more efficient in realizing this objective than a mitigation policy, the synergies should be exploited.

In order to reach a more precise definition of the term social capacity, we now elaborate on the connections between the social capacity base and relationships of substitution and complementarity. Table 4.2 lists a number of factors contributing to the social capacity base and the formal and informal institutions through which these factors are influenced.

Table 4.2 *The social capacity base – contribution of informal and formal institutions*

Factors contributing to the social capacity base	Informal institutions	Formal institutions
Information: Markets, prices, product innovations State performance Values, opinions, beliefs	Families, friends, peers, neighbours, kin, ethnic and work-related networks, informal groups, civil society groups, story-telling, religious activities	Newspapers, journals, magazines, books, radios, television, internet, freedom of the press, information disclosure laws, regulations Provision of infrastructure; roads, post offices, electricity, telephones Schools, school curriculum Political competition, citizenship rights, freedom to associate, civil society participation and other accountability mechanisms
Trust	Norms, values, interpersonal relations, social sanctions	Rule of law, independent judiciary, equity before law, enforceable human, consumer and property rights and contracts at low cost Secure financial institutions Schools, school curriculum, educational institutions Citizen participation in state activities and other accountability mechanisms
Credit	Ethnic and civil society groups, kin networks, friends, money lender, revolving credit societies	Banks, credit rating bureaus, training and marketing
Risk management: Health Livelihood	Differential healthcare spending, selective feeding Within family diversification; kin, social and work-related networks; intergenerational transfers; civil society groups	Health insurance policies, subsidized healthcare for the poor, education Investment in human capability, unemployment insurance, social security, safety-net schemes, pensions; laws enforced for open and fair hiring; investment incentives for private sector
Local public goods, basic services and common property resources	Community groups and committees	Coproduction with local groups through direct committees or indirect representation
Dispute management and resolution	Family, extended kin, ethnic network, traditional council, church, community groups	Lawyers, courts, tribunals, dispute resolution bodies, small court claims, police, social workers
Security	Norms, self-policing Neighbourhood or social group-based security systems	Police, armed forces, security personnel, law enforcement and fairness in law enforcement, investment incentives to private sector

Source: Based on Narayan (1999)

From Table 4.2 it is seen that the factors contributing to the social capacity base are simultaneously positively related to well-being and the functioning of the markets.

In the case of the markets, this can be illustrated by considering the introduction of a new technology. The market for a given technology will depend, of course, on the price and characteristics, such as efficiency potential compared with existing technologies, reliability, the lifetime of the product, and so on. However, the widespread use of the technology in question also depends on the degree to which information about the technology is spread through various informal and formal institutions and the existence of infrastructure – for example, in the form of roads and electricity. Similarly, the available financing options, the skills acquired through educational institutions, the nature of the legal system, and so on, will determine the adaptation of a given technology.

At the abstract level, the relation between social capacity and well-being can be illustrated as follows. The presence of several institutions, cross-group linkages and (preferably) complementarity represent an augmentation of the 'room for manoeuvre', both at the individual, group and society level because, at each of these levels, the choices available and the freedom to choose between alternatives are affected by the institutions present and their interaction. At the individual level an example is the choice of profession, which is assumed to affect well-being through a variety of factors such as income, self-realization, acquaintances, nutrition, etc. In the absence of cross-group linkages and a formal institutional setting, the individual will have very few options regarding choice of profession. In this case, the livelihood is most probably determined by the initial conditions – ie, the social status and connections of the family that the individual is a member of. The cross-group linkages provide a source of economic and social mobility – for instance, through access to credit schemes, risk management arrangements and community-based healthcare and schools. The formal institutions further expand the choices available to the individual through the provision of education, social security and insurance schemes of various kinds, alternative means of credit, enforcement of human rights, rules and regulations related to working conditions, and so on. Furthermore, they may increase the freedom of choice facing the individual by separating the provision of these factors from the norms, ethics and sanctions available to informal institutions based on kin and family.

The comments to Table 4.2 also correspond well with the previous suggestions regarding increasing the scale of transactions in the process of development and the necessary conditions for enabling this process. When the interaction between the formal and informal institutions is characterized by substitution or competition in one or more areas, the full potential of the corresponding factors contributing to the social capacity base cannot be realized, which reduces the scale of transactions and increases transaction costs as well as uncertainty.

More generally, the underlying assumption of the framework is that the social capacity is related to the ability and freedom of society to:

• adapt to new situations, including undertaking new activities; and
• utilize existing resources efficiently

The term 'social capacity' is consequently closely related to Sen's notion of capabilities, although Sen (1992) is largely concerned with individual capabilities and the aggregation of individual capabilities. The capability approach emphasizes that there are differences in what people can be and do, and that development consequently is closely related to freedom.[19] Central to the approach is the notion of functionings, defined as 'beings and doings' (Sen, 1992, p39). They cover a wide spectrum and examples include being healthy, getting adequate nutrition, participating in social activities and having self-respect. The difference between capability and functioning is that while a functioning can be regarded as one element, capability refers to a set of such elements, reflecting what a person can be or do, and thereby also reflecting the limitations of the possibilities facing that person. In this way, the two concepts can be said to address different dimensions of the development aspect: the functionings achieved are related to the achievement of well-being and the capability to function is related to the freedom to achieve well-being.

In the same way, social capacity is related to the freedom to achieve well-being and variations in what societies can be or do are reflected in differences in social capacity. Furthermore, social capacity refers to a set of elements, each of which affects the possibilities for the achievement of well-being. At the society level, examples corresponding to the first two functionings listed above are the provision of public healthcare and social security schemes. Elements that affect the remaining functionings listed above include the freedom to associate, the provision of infrastructure, information, civil society participation, education, the enforcement of human rights, and so on.

SOCIAL CAPITAL AS PART OF THE NATIONAL CAPACITY FOR GHG EMISSION REDUCTION

One approach to the assessment of GHG emissions and the potential for and costs of reducing these is to see them as depending on man-made, human, natural and social capital. In the baseline case, these four types of capital determine the structure of GHG emission sources and the GHG intensity per unit of activity for various sectoral activities and GHG emission sources.

This is illustrated by the following examples:

- Man-made capital: production technologies in manufacture, the energy system, land-use sectors, and so on, determine GHG emissions through production processes and efficiency
- Human capital: the educational level and professional skills of the population influence GHG intensity per unit of activity
- Natural capital: Available energy resources, soil quality, and so on, are resource inputs to production systems and determine sequestration capacity
- Social capital: determines efficiency of exchange processes, natural resource management, and establish a capacity for implementing GHG emission-reduction policies

Cost analysis, to which we will return below, often describes man-made and natural capital as driving forces in GHG emission scenarios and as the basis for more general development projections.[20] Typical areas included in mitigation studies have been GDP growth and other macroeconomic indicators, sectoral targets, detailed assessment of energy systems, fuels and emissions. Furthermore, GHG emission inventories have included the detailed assessment of physical sources and sinks. The studies have given very limited attention to the assessment of social capital aspects of mitigation analysis.

The baseline scenarios establish the starting point for policy implementation. The previous parts of the chapter have emphasized that policy implementation needs to take account of social capital aspects in order to be successful. It will be appropriate, therefore, to include a broad qualitative assessment based on the suggested social capacity framework in the national baseline scenario background. The assessment will indicate which of the quadrants in the framework that best describe the conditions for policy implementation in a given situation and it should be accompanied by a description of specific areas relevant to the baseline scenario. These areas may include:

- Existing information-sharing systems of markets and (other) institutions related to main GHG-emitting sectors
- Property rights, trust and enforcement related to the main GHG-emitting sectors
- Established government regulation applied to sectors with main GHG emissions
- Educational systems
- Characteristics of the markets related to main GHG emissions – for example, the number of companies, information flows and informal sector activities. This gives an indication of the thickness of the existing markets
- Established capacity for the management of complicated technical systems. This may be described in relation to existing advanced production processes or already operating energy technologies – for example are they managed by local experts or by non-domestic experts and companies?
- The availability of highly skilled technical staff in different sectors – for example, manufacture, energy and land-use sectors

In practice, it is most likely only possible to describe the above-mentioned social capital factors by a limited number of key statistical indicators.

In conclusion, the suggested approach is to make a parallel assessment of man-made, human, natural and social capital related to baseline GHG emissions leading up to conclusions on the general composition of national GHG emissions and on the social capacity related to the major emission areas. This conclusion should lead up to a subsequent analysis of the implementation aspects of GHG emission-reduction options.

IMPLEMENTATION OF GHG
EMISSION-REDUCTION OPTIONS

Different assessment tools such as conventional cost-benefit analysis (CBA), bottom-up studies and sector studies, assess the costs and benefits of implementing different mitigation options. In these standard approaches, it is more or less assumed that, apart from the costs of the resources involved, such as production factors, technologies, final products, overhead costs, and so on, no specific activities are necessary to promote project implementation. The implicit assumption is that the formal market or other appropriate institution establishes incentives for the agents to implement the project.[21] Implementation costs are thus only represented by the costs of planning activities, administration, information, training, monitoring and the like.

The successful implementation of climate change mitigation strategies will, however, typically necessitate measures involving costs that are additional to administration and training costs. Following the discussion at the beginning of the chapter, the existence of market imperfections, imperfect information, institutional failure, externalities, ill-defined and/or not well-enforced property rights, etc, indicate that implementation will generally not be pursued as a frictionless exchange process and that transaction costs may be significant. Put differently, unless a situation is characterized by conditions resembling quadrant 1 in the analytical framework presented earlier, additional measures for policy implementation are necessary.[22]

In the recent literature on climate change mitigation, these aspects are described as barriers to project implementation and the implication is that additional measures to remove – or alternatively reduce – these barriers are necessary in order to realize the desired outcome of a given project or strategy. The existence of barriers provides one explanation why the recognized potential for climate change mitigation with no cost or very low cost is not necessarily realized. The above examples of barriers, however, also indicate that barriers may very well be interrelated and rooted in the social organization of society. In other words, combinations of different aspects of social capital and the resulting social capacity may determine the barriers associated with implementation of a given GHG emission-reduction option. Adopting this perspective implies that the focus is moved from barriers to the underlying question of social capacity and from barrier-removal policies to the associated issue of social capacity-building.

Barrier Approach to Implementation

There is a growing interest in barriers to the implementation of climate change mitigation options[23] and they are frequently mentioned as an area in need of further investigation in project briefs.[24] As mentioned above, the interest may be seen as a response to the limitations of traditional cost-benefit analysis, bottom-up studies and sector studies.

While it is not the intention to provide a full account of the 'barrier approach', we will briefly outline a weakness of the approach and then turn to the 'social capacity' approach.

The barrier approach recognizes that a number of complicated issues arise in choosing barrier-removal policies due to intertwining of project, sector and macro-level with regard to transaction costs and policy instruments. Furthermore, the implementation of one policy may be in conflict with the implementation of another. Subsidizing electricity, for example, in the form of lifeline tariffs may, for instance, be in conflict with promotion of renewable energy technologies such as photovoltaics (PVs) as relative prices are distorted unless the renewable energy technology is subsidized too. Subsidies can also be a significant burden on government budgets and in this way conflict with the implementation of other policies. In the light of the above, it is recommended that the decision of which barrier removal measures to launch should be based on an assessment of the overall structure and functioning of the economy, including the policy and regulation framework. Such an assessment is necessary in order to obtain consistency and adequacy of the measures taken at different policy levels. Furthermore, it facilitates integration of the barrier-removal measures into the existing policy framework.

This assessment is, however, as a general rule based on a categorization of barriers according to type that, to a large degree, resembles a list of deviations from perfect market conditions. Below is a list of possible barriers to implementation sorted into different categories:

Market barriers:

- Markets tend to be missing or segmented
- Monopoly and/or monopsony
- Entry barriers
- Externalities and public goods
- Price distortions and the absence of price signals

Inflexibility and constraints of established technical systems:

- Capital irreversibility (turnover, infrastructure)
- Specific technology innovation and learning aspects (inertia, market development needs)
- Economies of scale

Institutional barriers:

- Unreliability or the absence of legal institutions → failure of enforcement systems, ill-defined property rights
- Underdeveloped financial markets → weak institutions, risk, limited supply
- Limited flow of information → inefficiency, inertia and collective action problems
- Administrative capacity constraints → failure to implement legislated policies

Human capacity barriers:

- Insufficient education level and coverage
- Limited supply of skilled labour and professionals

The inventory of barriers specific to the country in question forms the basis for decisions regarding barrier-removal policies.

The basic nature and purpose of a barrier-removal policy is, with reference to the bullet points above, to enhance market forces and private incentives to implement projects. A given policy can include a large number of different activities, including price incentives, information and the establishment of an institutional framework for market competition and access (such as capital market structure and institutions).

The policies are generally targeted to reduce transaction costs and intervention can be at the macroeconomic, sectoral or project level. Examples of barrier-removal options are: temporary support to the establishment of new companies, training programmes for new entrepreneurs, information campaigns, afforestation programmes and land reforms.

While this approach is perfectly legitimate – and in many respects resembles the approach suggested below – it suffers from a weakness. It addresses the limitations of the neoclassical approach using a neoclassical framework, focusing on the successive removal of individual barriers with little attention being paid to the social organization of society. The basic problem is that a coherent framework for assessing the overall functioning of the economy is missing. A listing of categories of possible barriers does not facilitate an assessment of whether the barriers are interrelated and rooted in the social organization of society, including the relations between and relative importance of the formal and informal institutions.[25]

Moving the focus from barriers to social capacity implies emphasizing the role of social organization in relation to implementation aspects. The assessment of the functioning of the economy can then be based on the analytical framework developed earlier in the chapter.

Social Capacity Approach to Implementation

The implementation analysis of GHG emission-reduction options should be carried out accordingly at two levels to complement a more traditional economic analysis:

1 Assessment of the existing social capacity for implementing a given project.
2 Analysis of the additional social capacity building efforts needed for the implementation of the project, including an assessment of the feasibility of undertaking these efforts.

The assessment of social capacity mentioned under 1 should be based on the analytical framework developed in the previous parts of the chapter. In the section below, the framework is further elaborated to provide a basis for such an assessment. The social-capacity assessment forms the basis for analysing which additional social-capacity building efforts will be necessary for the implementation of a given project. This analysis should result in a list of specific social capacity-building efforts to be undertaken to implement the given project. The efforts should include a qualitative description, estimate programme costs[26] and suggest a timeframe for the termination of efforts.

Social capacity-building efforts are defined as activities that enhance the capacity of individuals, institutions and society to undertake an activity or project that is in accordance with general economic policies and development programmes. Social capacity-building efforts are therefore related to general structural policies specifically when these facilitate exchange processes and thereby reduce transaction costs. In this way, it is expected that the capacity-building efforts are catalysts for the market development and commercialization of the specific options under consideration.

As a consequence, all the social capacity-building efforts are assumed to be transitory and evolving activities that establish a permanent capacity for a given option. Furthermore, this implies that the need for such additional efforts should be expected to diminish over time, depending on, among other things, the level of institutional integration, technological access and structure, and the effect of learning processes, as well as the degree of market imperfections, information, risk and uncertainty. It also implies that the requirement for social capacity-building efforts can be expected to be particularly high during a process of transition from a low to a high volume of transactions.

The implementation analysis in general will be an integral part of a more broad process of policy decisions. In such a process, the implementation analysis may be preceded by the assignment of priorities to a given portfolio of projects and/or strategies, and followed by a check for conflicts or inconsistencies with respect to existing projects and policies.

Elaboration of the Analytical Framework – the Quadrants

As will be recalled, the different enabling environments indicated by the analytical framework reflect variations in the social capacity for the implementation of policy options at a given level of society. This means that while an economy may be characterized by, for example, substitution and bridging social capital at the overall level, it is possible that the enabling environment for the implementation of a specific option is characterized by, for example, substitution and bonding social capital. In the same vein, options at sectoral and macrolevels can be linked with more than one type of social capacity.

The quadrants of the analytical framework are now elaborated on in turn. The findings of the framework are synthesized and examples of GHG emission-reduction options are given. The discussion is finally summarized in Table 4.3 (see p107).

Complementarity and Bridging Social Capital (Quadrant 1)

This is the ideal scenario, where the enabling environment is characterized by very favourable, earlier termed optimal, conditions and social capacity accordingly is high. The interaction between the formal and informal institutions is characterized by complementarity, contributing in a positive manner to all the elements of the social capacity base as illustrated in Table 4.2. Consequently, the ability to adapt to new situations and to utilize existing resources efficiently is high.

From a policy perspective, the markets work well and incentives are not distorted, which implies that minimum additional intervention and social capacity-building is necessary for the implementation of different options. Furthermore, the social capacity supports the implementation of all types of options. Under these conditions, models and results from mainstream economics are widely applicable, as the underlying assumptions are generally not violated.

However, in practice the degree to which mainstream economics is readily applicable will depend, among other things, on the level of distortions introduced by rules and regulations and whether property rights are assigned to all goods and services so that they can be traded in markets. Possible trade-offs with respect to existing policies and projects should therefore be considered.[27]

To exemplify this, consider the implementation of a carbon tax on end-use electricity consumption leading to increased energy efficiency through the effects on consumers' incentives to invest in energy-efficient lightbulbs, electrical appliances and energy-efficient behaviour such as minimizing stand-by time, turning lights off, and so on. Assume further that redistribution through a progressive income tax structure is a policy priority. In this case, the implementation of a carbon tax achieves the objective of increased energy efficiency but may create distributional concerns. Regardless of the increased energy efficiency, consumption at the household level will be determined partly by household characteristics, such as family size and the number of children in the household. As a result, the carbon tax may be regressive and imply a trade-off with respect to income distribution policies.

According to theory, the first-best policy response in this case is to take care of the distributional impacts through the tax structure and leave the carbon tax unchanged. Recycling the revenue from the carbon tax through the tax structure is one way of doing this.[28] If this is not possible, it is necessary to modify the mitigation policy if the distributional impacts of the carbon tax are to be avoided, although this implies a trade-off with respect to efficiency. One way of modifying the carbon tax would be to provide some energy at a 'lifeline rate', enabling low-income households to consume for essential needs.

Complementarity and Bonding Social Capital (Quadrant 2)

In this scenario, the formal institutional structure, including the market-supporting institutions, are in place but the intermediary link in the form of networks based on cross-group linkages is missing. As discussed at the beginning of the section, the costs of bonding social capital may prevail in the relative absence of cross-group linkages, as social and economic mobility is constrained and social cohesion at society level is missing.

The situation can be characterized as top-down development without bottom-up response and in this case well-being, social capacity and the functioning of markets is constrained by the absence of bridging social capital. This characterization may be most appropriate for specific parts of an economy, such as particular geographical locations (such as isolated communities or regions), certain areas of activity (such as fuelwood collecting or property rights to land), or a combination of the two. In the literature, these conditions are sometimes also taken to describe situations in which dominant groups based on bonding social capital have taken over or disproportionately influenced well-functioning governments.[29]

In this case, a role for general development policies is the development and maintenance of cross-group linkages to support economic and social mobility, and to secure social cohesion at society level. Efforts in these areas support a move towards quadrant 1 and may include facilitating information flows, ensuring civil society participation, providing adequate infrastructure and setting up educational programmes. Furthermore, it is noted that in the absence of bottom-up responses to policy options, the success of such options will rely on their continued implementation and financing.

At the specific policy level, co-production (or co-management) can facilitate the development of cross-group linkages and serve development purposes as well as climate change mitigation purposes simultaneously. An example of this is reported in a study related to forest management carried out in Gujarat in India (Pathan, Arul and Poffenberger, 1993). During the 1980s, an average of 18,000 offences was recorded annually, of which approximately 10,000 were cases of timber theft. Twenty forestry officials were killed and many more were assaulted in confrontations with communities and offenders. To change this bleak record, the conservator initiated an experiment involving joint management with communities. This included the widely publicized creation of forest protection committees, community meetings, and profit sharing of 25 per cent of timber returns with local groups. As a result, conflicts between officials and community groups diminished, community groups assumed responsibility for patrolling forests, reforestation was taking place and the productivity of land as well as returns to communities increased sharply.

Under the broad category of end-use efficiency options, electricity-saving programmes, the ILUMEX project is an interesting case example. The ILUMEX project was a programme for replacing 1.7 million ordinary lightbulbs with compact fluorescent lamps (CFLs) in two Mexican cities with an expected CO_2 reduction of 700,000 tonnes. As will be seen, this programme illustrates the point made under quadrant 1 that there are possible trade-offs with respect to other projects and existing policies to be considered when deciding on a given option. The cost-effectiveness of the programme may have been affected negatively by an existing subsidy on the price of electricity for poor households combined with targeting the CFLs to these households through subsidizing the price. Furthermore, the programme was implemented by the national electric utility, which may have dampened the effect on market development at the retail level (Martinot and Borg, 1998).

The ILUMEX project included a very detailed implementation strategy (Sathaye et al, 1994). The project is specifically targeted to low-income households because these households currently pay a low subsidized electricity price. Households with higher incomes pay a relatively high electricity price and savings by these consumers are therefore less profitable to the utilities. The CFLs were offered at a low subsidized price and their marketing included a number of options such as direct sales from utility offices, financing via the electricity bill, and 'mobile' information offices and sales campaigns in neighbourhood outreach offices. The programme also included a flexible monitoring system with feedback on incentives: lamp price, mobile sales offices, direct sales and installation force.

A number of expected implementation cost elements were assessed in ILUMEX. The direct subsidized purchasing price of the CFL was set at US$10. The basis for fixing this price was an assumed pay-back time of two years for consumers. The costs of programme execution and evaluation were assessed to be $1.64 per lamp, increasing up to $2.02 per lamp if all lamps had to be sold by mobile offices.

The ILUMEX project implies a number of interesting implementation policy issues. The first obvious question is whether it is desirable to target the programme to low-income households. These customers must be expected to have a relatively low capacity and incentive to react, reflected in short-time horizons (implying low pay-back time of the bulbs) and low information. Only the general electricity subsidies to this group suggest that a special effort should be made to increase electricity savings by low-income households. It is therefore a big question whether it would not be more efficient to suggest a deregulation of the subsidies for this group and eventually to support these households in another way. Another related question is the sustainability of the ILUMEX programme when special grants and marketing programmes have ended. It is difficult to see the long-term incentives of low-income households to buy new CFLs. Studies show that it is the more wealthy consumers who are leaders in technology adoption, due to factors such as higher marginal electricity rates, higher ability to pay and knowledge (Martinot and Borg, 1998). The cost-effectiveness and economic benefits of the project also appear to have been below the original forecasts for several reasons. In practice, consumers with high monthly electricity consumption purchased a large share of the subsidized CFLs. As these consumers pay electricity rates that are higher than the utility's marginal generation costs, the economic benefits were lowered due to lost profits from these consumers.[30] Other reasons include that the average lamp usage per day was 20 per cent lower than originally estimated and that the fuel mix for electricity generation has changed partly to gas.

The ILUMEX programme is therefore very illustrative of the interlinkages between market instruments and other barrier-removal policies. The exclusion of electricity subsidy removal in the implementation policies must be expected to lead to increasing short- and long-term programme risks and implementation costs, and it would therefore be recommendable to see the ILUMEX programme in a more general sectoral policy context.

Finally, policies aimed at the replacement of, or significantly influencing, existing informal exchange arrangements based on intragroup linkages should take the possibility of negative externalities into account. Interlinkages between markets and contracts need to be considered if policies are to succeed and if people are not going to be negatively influenced by the policies. Furthermore, options at the sectoral or macrolevel may entail less additional intervention for policy implementation than options at the microlevel. All options should take the incentive implications of the lack of bridging social capital into account. For instance, as a general rule signals concerning price and characteristics of a given technology will not be transmitted automatically, if cross-group linkages are absent. Implementation should secure that channels are provided to transmit such signals.

Substitution and Bonding Social Capital (Quadrant 3)

Like the first scenario, this is an extreme case implying an atomistic society, as would be the case in civil war or in the very early stages of development, before the emergence of a formal state. In this case, transactions are on a minimal scale and limited to small communities. Social and economic mobility is greatly constrained and formal markets do not function.

Under these circumstances, the existing social capacity for the implementation of general policies and GHG emission-reduction options is minimal. This can create something of a paradox in that it may only be feasible to implement specific types of options that do not rely on linkages for the transmission of incentives. The social capacity-building efforts associated with other options are likely to be too comprehensive to involve very long time-horizons, and to be associated with profound risks and uncertainties, making them generally unrealizable. The paradox arises because one of the factors needed for a first step towards development may be precisely the development of linkages.

In the absence of cross-group linkages and a well-functioning state, the community level must be assumed to be the best functioning level of society, and development policies as well as GHG emission-reduction options may be aimed accordingly at this level. As transactions are on a very limited scale, it is also noted that while production may be GHG emission intensive per unit of output, the total GHG emissions must be assumed to be relatively moderate.

A possible GHG emission-reduction option under these circumstances is the introduction of energy-efficient cook-stoves. Local craftsmen can produce such cook-stoves and the need for inputs is modest. The implementation of such an option on a large scale would imply, under the given circumstances, repeated implementation efforts in various communities, as markets and cross-group linkages are not present to facilitate the spread of the technology. This further implies that the implementation of the option is likely to be associated with high costs. The implementation strategy however, may include efforts to develop cross-group linkages between communities to increase information about the technology and to facilitate the acquisition of the skills needed to produce the cook-stoves. In this case, implementation costs can be reduced and the creating of cross-group linkages may serve general development purposes.

Substitution and Bridging Social Capital (Quadrant 4)

The enabling environment of this quadrant can be described as bottom-up development without top-down support. The presence of cross-group linkages facilitates a certain degree of social and economic mobility, as well as an increase in the volume of transactions compared with a situation in which reliance on intragroup linkages is dominant. However, in the absence of a competent set of formal institutions, the informal institutions act as substitutes, with resulting reliance on informal credit, barter economics, informal contractual arrangements, community-run schools and healthcare clinics, and so on. This has negative implications for the level of transactions, as transaction costs are high and the contributions to the elements of the social capacity base as presented in Figure 4.1, are confined to the informal institutions. In this way, constraints for the development of markets are imposed. Put differently, the available choices and the freedom to choose between alternatives are limited, with negative implications for social capacity and well-being at both the individual, group and society level.

As suggested by the previous parts of the chapter, these conditions may offer an adequate description of many developing countries and the quadrant will subsequently be discussed in more length.

From a policy perspective, there are several implications of the social capacity associated with this quadrant. First, development policies should focus on the missing state support. One aspect of this is the provision of an overall formal institutional structure related to areas such as credit markets, educational systems, infrastructure, information and insurance schemes. Another aspect is the relationship between state and society. Creating links between agents in the formal and informal institutions seems necessary for the success of the just mentioned initiatives. As mentioned in the discussion of quadrant 2, coproduction is one way of forging such links.

Second, as is the case in all the quadrants, the implementation of specific options needs to take account of the prevailing social capacity. Community or microlevel projects are less likely to be in conflict with the social capacity than projects at the sectoral and macrolevel. The full implementation of the latter will typically presuppose the existence of incentives mediated through formal markets and institutions that may not be in place. In this case, implementation should carefully assess the existing incentive structure and basis for transactions and consider any additional measures necessary for policy implementation. Further, to avoid futile competition between formal and informal institutions, it may be appropriate to concentrate on offering commodities or services for which networks do not provide close substitutes if coproduction is not feasible. For example, a policy option for the provision of energy services – for example, in the form of PVs or electricity grid extension – is a service for which networks do not provide a close substitute. By contrast, networks provide close substitutes for an option such as the introduction of energy-efficient cook-stoves. We return to this below.

The point about close substitutes basically reflects the possible consequences for markets of the presence of cross-group linkages in the absence of a well-functioning state:

- In some cases networks can prevent markets from functioning well. They may even crowd out markets, regardless of whether these are more efficient, and offer a wider range of products
- This is most likely to happen in situations where the informal institutions are involved in the production and exchange of marketable goods and services
- A corollary of the above is that while the informal arrangements may arise to reduce transaction costs, they may sustain or increase these costs in the longer run by posing obstacles to the emergence or development of formal market-supporting institutions
- Vertical integration is one way of mitigating a poorly developed institutional environment. But vertical integration can distort competition and facilitate monopsony situations

The broad relevance of these points still remains to be demonstrated empirically, but, as will be recalled, the example previously on the Ghanaian charcoal cycle supported some of the market consequences referred to above. Another type of GHG emission-reduction option that possibly illustrates the above points is the introduction of energy-efficient cook-stoves, which was also touched on in the discussion of quadrant 3. Characteristic of this option is that energy-efficient cook-stoves are a close substitute to existing cook-stoves and that informal institutions and markets are already involved in the production and exchange of the good.

In most strategy papers energy-efficient cook-stoves are mentioned as an opportunity with great potential for reductions of GHG emissions, as well as for contributing to development objectives through increased well-being for end-users. The latter may result from time saved with respect to collecting fuelwood and cooking, positive health impacts, costs saved in the cases where charcoal or fuelwood is acquired in the marketplace, and so on. In practice, however, implementation is often significantly constrained by several factors. Some of the factors commonly mentioned in country studies are the lack of skills of local craftsmen for the production of the stoves and the lack of demand. The latter can reflect a number of factors, including missing information about the product and its performance, lack of financing possibilities as the energy-efficient are typically more expensive than existing stoves but also have a longer lifetime, habits and preferences, and lack of supply. The lack of demand may also reflect that the designers of the stoves do not know the detailed problems of using them and cooking with them, and are not aware of what the users have evolved as a modus operandi. The users should be engaged in the design stage to avoid problems of adaptation. Furthermore, it has been questioned whether the estimated energy savings hold in practice.

The findings from country studies suggest that implementation should be accompanied by social capacity-building efforts facilitating the acquisition of skills, information flows and financing mechanisms. Furthermore, implementation should aim at using the existing markets for the distribution of the technology. In this way, the possible formal–informal market competition can be minimized.

At a more general level, the example is an illustration of an area where a so-called market transformation approach can be warranted. Such an approach provides a simultaneous market push and market pull for a specific technology and has been applied by the Global Environment Facility (GEF) to projects in the areas of efficient lighting, industrial boilers and household refrigerators in a number of countries.[31] To begin with, the enterprises are unwilling to produce because no established market exists and consumers do not demand the product because there is no familiarity with it. Market research, information, technology promotion and technical assistance can stimulate the demand for energy-efficient products and at the same time increase the producer's willingness to supply the product. Market transformation as described here often includes revised product designs, product labelling, appliance standards and consumer education.

Market transformation seeks to affect one or more of the following defining dimensions of a market:

- The number and nature of participants
- The variety and characteristics of the goods and services available
- The rules governing exchange in the marketplace[32]

In the above example, market transformation would be aimed mainly at affecting the variety and characteristics of the goods and services available. As has been discussed throughout the chapter, increasing the number and nature of participants can have various implications in different phases of a market development process and will generally be accompanied by changes in the rules for exchange. More participants will generally be expected to strengthen market competition, but may at the same time increase total transaction costs and the likelihood of failure of cooperation. This points to a possible role of public regulation or specific support activities, especially in the earlier market development states. The GEF and United Nations Development Programme (UNDP) suggest such temporary support activities in key strategy papers. It is argued that a number of advanced renewable energy technologies can be developed to be competitive energy production technologies with very low or zero GHG emissions, if they are temporarily supported. The technologies considered include wind turbines, PVs, biogas and biomass integrated gasifiers.

The arguments rely on expected learning curves, where costs decline with the increasing number of implemented plants. This cost 'gain' is expected to emerge as a consequence of enhanced markets, the development of local suppliers of technology components, and management experiences gained. The recommended implementation policy is then to provide the temporary support needed to ensure an implementation scale where costs are driven down to the competitive level. The same type of recommendations can be found in relation to the discussion of infant industries, for example. Here, the arguments build on the economic concepts of the economies of scale and market imperfections, resulting in a lack of knowledge and limited credit markets that prevent producers from borrowing against future falls in the costs of production.

The learning curves argument is specifically outlined in the following way by UNDP (1996, p82) in the case of biomass integrated gasifiers. It is expected that the costs of the biomass integrated gasifiers will follow a learning curve. More specifically, assumptions are being made concerning the costs of 10 units of a 25,000kW plant that is scheduled to commence operation in north-east Brazil in the late 1990s. Shell researchers (Elliot and Booth, 1993) estimate that costs will decrease by 20 per cent for each cumulative doubling of production. This will imply a total cost of learning of US$0.12 billion. It is difficult to judge how much of such learning costs (or gained benefits) can be transferred from one project to another, especially if the projects are implemented in different national settings. The development of indicators for the assessment of learning effects is therefore an interesting area of study.

Returning to options regarding end-use efficiency, many country studies for developing countries have identified energy-efficiency improvements related to industrial motors and boilers as attractive low-cost mitigation options (UNEP, 1994; Halsnaes, 1996). In these countries, the industry sector is often using inefficient and economically outdated production equipment as a consequence of, for instance, capital constraints and inertia. The introduction of new technologies may therefore at the same time save energy consumption and reduce GHG emission at low costs or in some cases even with a benefit. The challenge is to design social capacity-building policies that can help the implementation of the options.

Several social capacity-building policies can support the implementation of new efficient boilers and motors in industry. These include financing, information and training, and more specific market instruments such as tax and subsidy policies. The costs of specific financing options and information and training efforts can be measured relatively easily and integrated in the total cost assessment. The main uncertainty here is related to the 'efficiency' of the policies in getting the actors to implement the projects. The costs and consequences of using market instruments are more difficult to assess. Market instruments can include, for example, carbon taxes on fossil fuels, where part of the revenue is recycled to industry as investment grants for the boiler and motor options. The market instruments in this case have a number of general direct and indirect economic impacts, in addition to the incentive to implement new boilers and motors. These broader impacts can only be assessed at sector or macroeconomic level and the implementation cost assessment for these instruments must therefore be directly related to these more general assessments. Another implication of the very general character of the impacts of market instruments is that the introduction of such policies should be considered for a larger group of sector policies, and not for individual projects alone.

A number of country studies for developing countries have made a qualitative assessment of implementation policies in relation to the mitigation option efficiency improvements in industrial boilers and motors. A study for Venezuela (UNEP, 1994) assessed a large technical potential for boiler efficiency improvements in industry with very low costs. However, the implementation of new efficient boilers was difficult due to a number of factors. It was emphasized particularly that companies preferred to maintain established equipment because spare parts were easily accessible and because the technology was well known. It was assessed that the decision-makers in the companies had information about the new technologies, but that they underestimated the potential efficiency gains of implementing them. The high up-front installation cost of new boilers in relation to just maintaining the old ones was also seen as a significant implementation cost. On this background, it was recommended to design an implementation strategy that included information campaigns, third-party financing or other financial mechanism, and general market creation activities where demonstration projects and the local supply of spare parts for the new equipment should be subcomponents.

To sum up, the discussion of the quadrants has aimed at illustrating the different enabling conditions for the implementation of different options reflected in the four main types of social capacity. Table 4.3 below summarizes the primary characteristics and policy implications of the quadrants.

Elaboration of the Analytical Framework – a Portfolio of Projects for a Number of Countries

So far, the implementation analysis has been described in terms of the implementation of a single option. The analysis based on the analytical framework

Table 4.3 *Social capacity for the implementation of policy options*

Prevalence of	Complementarity	Substitution
Bridging social capital	Transaction costs are low, information flows high and the formal and informal institutions reinforce each other in the provision of the elements contributing to the social capacity base Policy implications: Implementation requires a minimum of measures additional to standard assumptions The social capacity supports all types of projects Policies, however, should take account of possible trade-offs with respect to other policies and projects	Transaction costs are high and formal markets are not well functioning. Networks are substitutes for formal markets. Networks may crowd out markets even when these are more efficient and offer a wider range of products Policy implications: Focus on the missing state support (credit markets, education, information) and forge links between state and society Implementation requires measures additional to standard assumptions Offer commodities for which networks do not provide close substitutes Community level projects Coproduction Options may consider a market transformation approach Careful assessment of additional interventions for options at sector and macrolevel
Bonding social capital	Transactions are impeded as the intermediary link in the form of bridging social capital is lacking. Top-down development without bottom-up response, implying no evolution of social capacity. In principle it is necessary to continue policy implementation and financing indefinitely Policy implications: Efforts to build and develop cross-group linkages to support economic and social mobility, ease information flows, ensure civil society participation and secure social cohesion at society level Coproduction Take account of interlinkages Careful assessment of additional interventions for options at community level	Transactions are on a minimal scale and limited to small communities. Social and economic mobility is greatly constrained and formal markets are non-existing Policy implications: Implementation of specific types of options that do not rely on linkages for the transmission of incentives Community-level projects Policies should aim at facilitating the development of cross-group linkages Implementation will be associated with high costs

may also be carried out at an overall level, analysing the implications of differences in social capacity for implementation, considering a portfolio of projects for a number of countries. This exercise would provide general insights concerning the dependency of specific policies and projects on social capacity in relation to the main GHG emissions. Examples of this are given in points 1 to 6 below. In addition, the exercise would provide the basis for generalized findings concerning which social capacity-building efforts should be undertaken to implement different options under different conditions.

While it is beyond the ambition of this chapter to provide a generic analysis of a portfolio of options for a number of countries, the general comments and examples given in the discussion of the four quadrants permits some tentative suggestions regarding the output of such an exercise. These suggestions are added to the general examples of outputs of such an exercise listed below.

For each of the quadrants of the framework, the exercise will identify:

1   Options that are relatively independent of other activities:
   • The relative independence of other activities can be an inherent characteristic of a given option, rather than related to a specific type of social capacity. Options such as large, centrally managed power facilities belong to this category. At the general level, there will be a rationale for the implementation of this option in quadrants 1, 2 and 4.
   • Alternatively, independence may arise from the characteristics of the prevailing social capacity. In this case, independence in a given context reflects a general lack of linkages: across groups, between the state and the society, or both.[33] The lack of linkages needs to be taken into account for the successful implementation of a given option. In quadrant 3, the lack of linkages is generic, which indicates that implementation may be associated with considerable costs.
2   Options dependent on formal and (or) informal sector activities due to direct linkages or spillovers:
   • This generally applies to options related to the macrolevel, such as liberalization and privatization schemes, and environmental taxes aimed at increasing efficiency by changing the overall incentive structure. Successful implementation of these options presupposes that the relevant institutions, especially markets, are in place to transmit the incentives and that the government is capable of enforcing the policies. This suggests that, for example, environmental taxes are most suitable under the social capacity conditions of quadrant 1.
   • Options affecting interlinkages between contracts or institutions. Policies aimed at the replacement of, or significantly influencing, existing informal exchange arrangements based on intragroup linkages may create negative externalities. As discussed in the previous sections, this may apply, for example, to credit-labour interlinkages in rural areas.
   • Options where project implementation and sustainability beyond the implementation phase requires developing financing mechanisms. For example, if consumers do not have access to credit, the higher equipment cost of energy-efficient cook-stoves and other cooking appliances may prevent their usage, even when they are more cost-effective.

3   Options that will require radical changes in market structures or will imply significant redistribution of rents among stakeholders:
   • The first type generally applies to the options mentioned under 2, related to the macrolevel, such as liberalization and privatization schemes, and environmental taxes aimed at increasing efficiency by changing the overall incentive structure.
   • Examples of the second type of options include reduced coal consumption in a country with domestic mining or the closure of industry, as well as land reform programmes.
4   Projects related to activities that are not integrated in formal markets and may not be well governed:
   • Prominent examples are land-use sector activities. Implementation can require changes in property rights and other complicated reform programmes.
5   Options that are assessed to be in line with general development patterns and therefore only require a limited range of information and training activities:
   • This applies to all options in quadrant 1. In quadrant 2, options at the macro- or sectoral level, such as liberalizing energy prices or reform of the power sector, fall under this category. In quadrant 4, providing basic energy services at the community level is an example of such options.
6   Cases in which it must be concluded that a project cannot be implemented or will work very inefficiently if implemented because a number of prevailing incentives (including market prices) will work against it:
   • As indicated above, this applies to most options if they are implemented under the conditions of quadrant 3.

The points suggest that in most cases, successful implementation of different options will require some social capacity-building activities. We now turn to these activities and discuss how they should be measured in the context of projects.

Social Capacity-building Activities

A central theme of the chapter is that in most developing countries, the traditional assumptions in economics regarding efficient resource allocation and well-functioning markets basically do not reflect real conditions for policy implementation. A number of very specific assumptions must be applied therefore on how the implementation of mitigation options can be supported by specific social capacity-building activities.

At the beginning of the section, it was noted that social capacity-building involves activities that enhance the capacity of individuals, institutions and society to implement a given option. The activities should therefore support exchange processes related to project implementation and accordingly may be aimed at correcting market failures directly or at reducing the transaction costs in the public and/or the private sector. Broad (and interrelated) examples are improvement of the institutional capacity; reduction of risk and uncertainty; facilitating of market transactions; and enforcement of regulatory policies, and so on. In this way, it was also noted that the activities should establish a permanent capacity for a given option and that consequently they are assumed to be temporary. If the activities are undertaken successfully, the need for social capacity-building efforts is therefore expected to diminish over time.

With respect to the quadrants of the analytical framework, this implies that social capacity-building activities should facilitate a move towards quadrant 1. It also implies that the requirement for social capacity-building efforts can be expected to be particularly high during a process of transition from a low volume of economic transactions to a high volume of economic transactions. This was illustrated in the discussion of the quadrants above and provides a basis for describing the general types of social capacity-building that are needed for the implementation of options in the different quadrants. Table 4.4 lists examples of social capacity-building activities and the quadrants where they can be assumed to be most relevant.

The effects of the social capacity-building activities mentioned in Table 4.4 reach beyond the implementation of a specific option. Put differently, the activities are associated with positive externalities that are very difficult to estimate. The problem is that if the costs of the social capacity-building activities are debited fully to one project, then this project will never have the required cost-effectiveness unless the external benefits are also credited. This raises the question of how the activities should be measured in the context of projects.

One way is to consider the whole portfolio of GHG emission-reduction projects and to allocate a total cost, X, to the portfolio in support of social capacity-building. It is then possible to allocate this total cost to each project in proportion to the project's value relative to the portfolio. Assume that the total value of all GHG emission-reduction projects is V, and a project, I, has a value of $V(I)$. The share of the social capacity-building costs to be debited to project I is $X(I)$, where:

$$X(I) = X * (V(I)/V)$$

Such an allocation has the advantage of being transparent and easy to administer, but it is not entirely unproblematic. One obvious point is that *ceteris paribus* the larger the portfolio is, the smaller $X(I)$ will be. The allocation also has to take the issue of time into account. If the total cost in support of social capacity-building is decided for a portfolio of projects to be implemented within a given timeframe, projects starting late in that period may benefit from social capacity-building activities already undertaken by other GHG emission-reduction projects. Another issue is that the projects that are largest and thus contribute most to the total value of the project portfolio are not necessarily the ones that require most social capacity-building activities. For example, an option that is relatively independent of other activities, such as a large, centrally managed power mentioned previously, may have a large value in the project portfolio but will not necessarily entail much social capacity-building.

In order to reflect these aspects, some type of weighting can be introduced into the allocation of costs. These weights may reflect, for example, development policy priorities. A higher share of the total costs than implied by the equation above can accordingly be allocated to projects with associated social capacity-building activities that are considered to be of high development priority. The time aspects can also be addressed by making the share of the total costs to be

Table 4.4 *Examples of social capacity-building activities*

Social capacity-building elements	Examples of activities	Relevant quadrants
Market creation, possibly with public sector involvement in the transition period to stimulate demand and supply of a given technology	Temporary support to specific demonstration projects, market research, information, technical assistance, technology promotion	2, 3, 4
Establish and enforce property rights	Land reforms Public afforestation programmes	2, 4 1, 2, 3
Establish monitoring and enforcement systems	Reporting systems	2, 3, 4
Regulate competition by introducing more market actors	Information campaigns, soft loans to developers of renewable technologies, privatization of public electricity utilities	1, 2, (3), 4
Deepening financial markets to support efficiency in savings and investment decisions	Support financing mechanisms (eg GEF)	2, 4
Launch technical standards to be met in a given timeframe	Efficiency standards for electricity appliances	1, 2, 4
Price liberalization to support international competition and incentives for efficiency	Removal of price subsidies	1, 2, 4
Target inflexibility and constraints of established technical systems: Timing of infrastructure investments Subsidy to capital turnover projects Subsidized credit to support research, development and learning processes	 Long-term planning of power production and transmission Specific capital grants Demonstration and research programmes, training programmes	1, 2, 3, 4 1, 2, 4 2, 4
Coordination and integration of specific climate change mitigation efforts in general investment policies	Information, capital subsidies	1, 2, 4
Reduction of commercial and purchase risks	Codes, standards, and certification and testing agencies	2, 4
Institutional set-up for risk reduction and/or risk pooling	Deepening of formal financial markets, commercial financing and guarantees, revolving funds financing through specialized government agencies	2, 3, 4

Table 4.4 *Continued*

Social capacity-building elements	Examples of activities	Relevant quadrants
Establishment of specific institutions to reduce uncertainty and transmit information on opportunities, costs and benefits	NGOs, civil society groups, community groups, educational institutions, information centres, awareness campaigns	2, 3, 4
Establish international mechanisms for technology transfer	Clean technology mechanism, Clean Development Mechanism	2, 4
Improvement of decision-making processes	Training and education activities, civil society groups	2, 3, 4

allocated to one type of social capacity-building activities dependent on previous activities of the same type. Alternatively, all social capacity-building costs could be allocated to a central budget and none to the marginal valuation of a particular project. This type of allocation would also permit considerations of development priorities and time aspects.

COST CONCEPTS RELATED TO SOCIAL CAPITAL AND SOCIAL CAPACITY

Countries with given man-made, human and natural capital can have different social capacity for implementing GHG emission-reduction policies, reflecting differences in social capital and in the combinations of various forms of social capital. The differences in social capacity have implications for the baseline GHG emissions and for the relative size of the reduction potential to total emissions, as well as for the efficiency and costs of reduction policies. A low capacity for policy implementation will be reflected in high social capacity-building costs and long time horizons needed to establish a more permanent capacity.

Implementation costs are one of the most critical components in providing more reliable estimates of mitigation potential related to the aspect of policy recommendation in different settings. However, it is not trivial to develop a framework for assessing implementation costs that addresses the many specific national issues that will be critical in strategy development. We have emphasized that implementation costs cover all costs associated with the expected requirements to realize a project or sectoral strategy. As was indicated earlier, these costs can be seen as consisting of two separate elements. Let us call the first element administration costs. This element includes the costs of activities that are directly related and limited to the short-term implementation of the project or sectoral strategy, such as the costs of planning, training, administration, monitoring, and so on. Administration costs are thus the costs included in standard CBA, bottom-up and sector studies. The second element is the costs of additional social capacity-building activities needed for implementation. As illustrated in the earlier parts of the chapter, these costs are time and context specific and are incurred to enhance the social capacity for policy implementation and their effects are therefore not limited to the immediate project or strategy.

The inclusion of considerations of social capacity in implementation cost assessments requires that quantitative indicators and baseline scenarios be developed. The development of a comprehensive set of qualitative indicators will provide a valuable first step in this process.

It also means that implementation costs should be measured as social costs, because social capacity-building costs inherently are social.[34] They are incurred to reduce the social costs in the longer run by making regulation and policy instruments work. The measurement of social implementation costs reflects the general welfare perspectives of the policies.

One aspect related to the measurement of social capacity-building costs is that different combinations of social capacity-building activities can be suggested to implement projects or strategies, but the specific design and the costs of the different policies depend on the comprehensive social capacity-building effort. This can be illustrated in the case of electricity-saving efforts for private households. The policy can be implemented through information campaigns and technical standards, through market instruments like taxes or subsidies, or more realistically by a combination of the mentioned policies. The cost of each of the social capacity-building policies must be expected to decrease if they are combined with other policies – price signals will work better if households are well informed not the other way around. However, in practice social capacity-building policies may not be able to take full advantage of such synergies because other political or social considerations prevent the use of certain instruments. Furthermore, as noted previously, conflict between the instruments used at the different levels is possible, or even likely.

Indicators for Measuring the Positive Externalities of Social Capacity-building

Social capacity-building activities have different sorts of programme costs, as indicated above, but they also have a number of side-impacts that are particularly interesting to measure. In the following, it is outlined how cost concepts may be applied to measure these impacts:

1   Direct impacts: a social capacity-building component will result in decreased costs of running a specific project over time that does not reflect price changes in major inputs or outputs. The decrease in costs reflects that actors learn, efficiency is increased, markets are established where information is shared and transaction costs are reduced, and other actors participate in a more effective way (local manufactures, the supply of spare parts). All together the indicator here should reflect the total value of cost savings over time due to learning.
    Indicator: the total cost of the project over its lifetime, assuming that annual costs would be the same as in the first years of operation (assuming no learning), minus the total costs of the project assuming that the capacity-building activities are undertaken.
2   Indirect impacts on the costs of implementing similar policies or technologies: the focus here is on indirect cost savings for activities other than the project

in question, reflecting that an institutional structure, which also supports other activities, is established, making additional social capacity-building efforts cheaper for subsequent programmes.

Indicator: the total cost of the activity over the lifetime, assuming that annual costs are what they would have been without the previously implemented project activity (assuming no social capacity-building), minus the total costs of the project assuming that the social capacity-building activities have been undertaken in relation to the project.

3   Local business and market development impacts:

a   The aim is to reflect how a project affects the number of local stake-holders participating in a market, the distribution of costs and benefits among these (losers and gainers) and the cross-linkages created. It is assumed to be positive to get many local actors involved, if this is in line with the cost-effective management of the project and if it supports the development of more business activities. Two aspects should be considered in relation to indicators: the direct market intervention of the project and the market development resulting from the project.

Indicators: the number of participating companies and institutions over time, cross-membership of institutions over time, direct investments or sales supported by the project through financial flows to local implementing institutions or companies and to other local partners, indirect investments or sales not financed or subsidized by the project.

b   As a consequence of a, local business partners and institutions can improve their efficiency in relation to other activities and exchange processes, and in this way benefit from the capacity that is established according to a.

Indicators: the increase in the profitability of the other activities measured directly or as a function of codes, standards and the certification of products; market contract forms; development of prices, characteristics and quality of products, and the capabilities of the participants. The number of new activities in which the parties are involved as a consequence of the newly established capacity.

4   Direct impacts on the well-being of people who are related to the project. This can be measured as the impact on capabilities.

Indicators: access to energy services (for example, measured by coverage, reliability of supply and affordability), time, health, education, income distribution, participation in civil society activities and networks, and so on.

5   Indirect impacts on capabilities in a broader setting due to indirect economic impacts of the projects.

Indicators: access to energy services (for example measured by coverage, reliability of supply and affordability), time, health, education, income distribution, participation in civil society activities and networks, and so on.

Including Social Capacity-building Indicators in Project Assessment

In the following, we show how some of the types of impacts described above may be included in project assessment. The project assessment is based on a traditional cost-benefit analysis approach applied to studies of GHG emission-reduction policies and the examples are constructed using data from two of the Botswana case studies presented in Chapter 7 of this book.[35] The projects considered in this section are a road pavement project and an efficient industrial boiler project.

Three types of cost assessment are presented: gross financial, net financial and social costs. As described in Chapter 7, gross financial costs are the costs of running the GHG emission-reduction project excluding costs of the substituted baseline activity. Net financial costs are the financial cost of a given project measured as an incremental cost in relation to running a baseline case. In the case of efficient industrial boilers presented below, the gross financial costs of the project are the capital cost of the motor, the cost of operation and maintenance including power, and the implementation costs. The net financial costs for this project are the gross financial costs minus the savings in operation and maintenance cost including the value of electricity savings assessed in relation to the baseline case. Cost savings due to learning effects – leading to decreasing implementation costs over time – are also subtracted from the gross financial costs.

These assessments are supplemented with a social cost assessment. In the Botswana case studies, the direct cost assessment is supplemented with information about side-impacts on local air pollution, employment generation, and health impacts in the coal-mining sector.[36] To include other types of impacts as described in the previous section, additional assumptions are made concerning direct and indirect cost savings related to learning effects, market development effects, and direct and indirect effects on well-being through income distribution and time savings. Specific national data has been used to estimate local employment impacts, time-savings and income distribution effects, while international data has been used to represent damages of local air pollution and health impacts in the coal-mining sector. Furthermore, to represent the learning effects and market development effects, we have made assumptions regarding costs of information and training and changes in interest rates based on data from international studies.

This means that the side-impacts have not been accurately estimated and the social cost estimates presented below should therefore only be considered as an illustrative numerical example of the costs. However, the established information is considered to provide a number of useful insights about the consequences of including broader issues related to social capacity in project assessment.

Table 4.5 below gives a characterization of the Botswana case studies considered and the assumptions made on social capacity-building effects. Each of these effects and their estimation is described in more detail below and in Appendix 4.1, at the end of this chapter.

The implementation cost assumptions made for the analysis are as follows:

- To provide an example of the effects of local business and market development, we split the efficient industrial boiler example into two cases. In the first case local market and business development is low, which is reflected in high interest rates due to the lack of credit opportunities. We have applied a discount rate of 13 per cent in case 1 to mirror this
- In case 2 we assume that another project on cash crop production and marketing has been implemented. This project is totally unrelated to the GHG

Table 4.5 *Botswana case examples, inclusion of social capacity-building*

Case	Project example	Baseline case	Assumptions made to include social capacity-building and in adjusting financial costs to get social costs
Road pavement	Pavement of 531km sandy roads linking Botswana to Zambia, Zimbabwe and Namibia	Sandy roads with similar traffic level, but with 50% higher fuel consumption for the same traffic	The social costs include the benefits of reduced SO_2, NO_x and particulate emissions Increased employment in the construction phase Income distribution effects through employment generation Time savings for users of the road
Industrial boilers	Improved coal-fired boiler with 85% efficiency by installing economizer on traditional boilers	Case 1: Existing coal-fired boiler with 79% efficiency Case 2: Existing coal-fired boiler with 79% efficiency Previous implementation of project related to production and marketing of cash crops	The social costs include the benefits of reduced SO_2, NO_x and particulate emissions Health impacts implied by decreased coal production in the mines Direct and indirect learning effects are included by reducing implementation costs Effects of local business and market development are included in case 2 by using a lower discount rate reflecting the implementation of the cash crops project

emission-reduction option, but is associated with positive externalities, as local markets for credit have been developed as a consequence of the project. Using a discount rate of 10 per cent in case 2 reflects this effect

- The Botswana road pavement project is an example of an option that is characterized by being relatively independent of other activities. As described previously, the implementation of such options will generally not require extensive additional activities. Hence no additional implementation costs are included for that project. In contrast, the efficient industrial boiler project may require more additional implementation activities as implementation in this case depends on the functioning of markets, information flows, and so on. The extent of such activities and the corresponding implementation costs will depend on the social capacity base. We have assumed that implementation costs, including training and information activities, amount to 15 per cent of capital and operation and maintenance costs

Including these costs, as well as including an indicator for local market and business development, represents one way of adjusting the baseline case to reflect social-capacity aspects as suggested in the section on social capital as part of the national capacity for GHG emission reduction.

The estimates of the financial and social costs are presented in Table 4.6.[37] Table 4.6 shows a marked difference between gross financial costs, net financial costs and social costs. The difference between net financial costs and gross financial costs is particularly big for the road pavement project, where the value of the fuel savings included in the net financial costs is very large. This reflects that the project is assumed to substitute a very inefficient baseline activity. In the efficient boiler project, the difference between the two types of financial costs is also almost entirely due to fuel savings. The savings arising from learning effects in case 1 and 2 are US$0.06 and US$0.09 per tonne CO_2 reduction respectively. The negative net financial costs of the projects imply that they are justified solely on efficiency grounds and that they do not need support in the form of social cost assessment to justify them.

Social costs are lower than financial costs for both the projects considered. The efficient industrial boiler project has the largest benefits on local air pollution because this project substitutes pollution-intensive coal consumption. However, the road pavement project still remains more cost-effective than the efficient industrial boiler project. The benefits of employment and time-savings in the road pavement project are small relative to the benefits from reduced local air pollution. Among other things this reflects the capital-intensive nature of the project and the inefficiency of the baseline activity.

Both the above conclusions and the relative rankings of the projects given in Table 4.6 hold in the absence of implementation costs (see Table 7.5); hence in this case implementation costs do not change the broad conclusions on rankings. However, the addition of these costs does show some interesting points. First, comparing case 1 and 2 in Table 4.6, it is possible to assess the effects of including an indicator of local business and market development. Gross financial costs increase by about 11 per cent and net financial costs by about 22 per cent. Second, compared to the employment and time-savings effects, the effects of local business and market development are quite small (their inclusion changes social costs by 3.4 per cent).

The coal-mining impacts are so large that their inclusion in the social costs of the efficient industrial boiler project makes this project the most cost-effective as it substitutes coal consumption.

Trade-offs between Project Efficiency, Risk and Local Capacity-building

There will most likely be both trade-offs and synergies between local institutional capacity-building and the risk of projects. A project that depends only to a very limited extent on local institutions and capacity may have a low risk because it can be centrally managed, but may at the same time not have significant social capital impacts. On the other hand, a project with many local linkages can be very successful from a project, social capital and development perspective if the linkages are handled well, but the opposite if not.

Table 4.6 Financial and social costs of GHG emission-reduction projects in Botswana

costs	Gross financial costs	Net financial	Social costs impacts	Local air-pollution distribution	Employment impacts, income adjusted	Time savings coal-mining	Social costs including impacts	Coal-mining impacts
1 Efficient industrial boilers: case 1	6.1	-4.5	-55.3	-50.8	0	0	-	-
2 Efficient industrial boilers: case 2	5.4	-5.5	-57.2	-51.7	0	0	-221.8	-180.6
3 Paved roads	13.0	-101.2	-142.4	-38.3	-2.8	-0.1	-142.4	0

Note: All costs are in US$ per tonne CO_2 reduction. A 10 per cent discount rate is used for 2 and 3, while a 13 per cent discount rate is used for 1, reflecting the lack of local business and market development

A strategy that tries to optimize local linkages and thereby local capacity-building therefore should include:

• an assessment of risk;
• an assessment of intervention possibilities; and
• an evaluation strategy for assessing social capital linkages over time as part of the project implementation and operation, including the relations with parties that participate directly and indirectly in the operation and implementation of the project.

The above also suggests the following point. If implementation costs are not included in the assessment of social costs and if social and external benefits linkages are not evaluated, there is a risk that:

• the rankings of projects may be misleading, with projects that have high costs being favoured and then found to fail when inadequate attention is paid to these elements; and
• opportunities to undertake projects that have important development and social impacts will be missed.

CONCLUSIONS

GHG emission-reduction options are closely related to general development aspects because the options aim at increasing the efficiency with which existing resources are used and this is an inherent aspect of development. Efficiency improvements may thus arise from and depend on structural changes and the general improvement of the functioning of the economy, including the functioning of formal and informal institutions.

Two aspects of social capital have been considered in this chapter: linkages between and within groups, and the interaction between state and society. Focusing on the nature of relationships between and within groups and between the state and society, and drawing on insights from economic theory, a simple analytical framework for the assessment of social capacity was developed. The main conclusions to be drawn from the analytical framework are:

• The selected options, the implementation policies and the mechanisms for implementation must reflect the prevailing social capacity if implementation is to be successful
• The social capacity has implications for the implementation of different options. Some types of projects and strategies can only be implemented if the social capacity is strong
• The implementation of different options will in most cases necessitate social capacity-building activities. The relevant social capacity-building activities will depend on the social capacity in a given situation

The chapter has also argued that the conventional approach to barriers and barrier removal suffers from a weakness because it does not facilitate an assessment of whether the barriers are interrelated and rooted in the social organization of society. The basic problem is that a coherent framework for assessing the overall functioning of the economy is missing. The inclusion of social capital aspects permits the development of such a framework and provides a link between the implementation of GHG emission-reduction options and the concept of social capital.

The framework also facilitates the assessment of the need for social capacity-building efforts for the implementation of projects. Social capacity-building involves activities that enhance the capacity of individuals, institutions and society to undertake an activity or project that is in accordance with general economic policies and development programmes. In this way, it is expected that the capacity-building efforts are catalysts for the market development and commercialization of the specific options under consideration. As a consequence, social capacity-building efforts are assumed to be temporary activities that establish a permanent capacity for a given option. The chapter has outlined how the costs of such activities may be measured in the context of GHG emission-reduction projects.

Implementation costs are one of the most critical components in providing more reliable estimates of mitigation potential and related policies. It is, however, not 'trivial' to develop a framework for assessing implementation costs that addresses the many specific national issues that will be critical in strategy development. However, the chapter has presented some potentially useful indicators and examples of their inclusion in project assessment were given.

It was further pointed out that if implementation costs are not measured as social costs, there is a risk that the projects that seem best suited for implementation under given circumstances – because they are associated with the lowest implementation costs and risks – will be the projects that simultaneously have the least significant development impacts.

NOTES

1   For a review of the concept of social capital and its history, see, for instance, Woolcock (1998).
2   In the sociological, anthropological and political science literature, social capital relates to the norms, networks and organizations that influence policy formulation and decision-making, and through which people gain access to power and resources.
3   See, for instance, Woolcock and Narayan (2000) and Evans (1996) for details.
4   See, for instance, Martinot and McDoom (2000).
5   See, for instance, World Bank (1996, 1997a) and Martinot and McDoom (2000).
6   For this purpose, the definition of institutions captures the relevant aspects of social capital. To elaborate on this, consider the following definitions of social capital and institutions respectively.
    Social capital:
    Narayan and Pritchett (1997, p2): '. . . the quantity and quality of associational life and related social norms.'
    Coleman (1990, p302): 'Social capital is defined by its function. It is not a single entity, but a variety of different entities having two characteristics in common: They all consist of some aspect of social structure, and they facilitate certain actions of individuals who are within the structure. Like other forms of capital, social capital is productive, making possible the achievement of certain ends that would not be achievable in its absence.'

Portes (1995, p12): 'Social capital refers to the capacity of individuals to command scarce resources by virtue of their membership in networks or broader social structures.'

Loury (1992, p100): '. . . naturally occurring social relationships among persons which promote or assist the acquisition of skills and traits valued in the marketplace.'

Narayan (1999, p6) '. . . the norms and social structures of society that enable people to co-ordinate action and to achieve desired goals.'

Institutions:

Halsnaes et al (1998, p63): '. . . forums with implications for rules of interaction. . . Examples of institutions are families, "social rules", markets, etc.'

Todaro, M P (1997, p699): 'Norms, rules of conduct and generally accepted ways of doing things.'

7 See, for instance, Portes and Landolt (1996).

8 See, for instance, Putnam, Leornardi and Nanetti (1993) and Rubio (1997).

9 The notions of bonding and bridging social capital are introduced by Gittel and Vidal (1998), building on Putnam, Leornardi and Nanetti (1993) and used by Narayan (1999) and Woolcock and Narayan (2000). Most authors choose to relate the ties or linkages to the community level rather than the group level. We apply the notion of linkages to the group level to take account of the fact that cross-group linkages may reduce inequality within communities as well as across communities.

10 Small defined as 1–30 employees.

11 We return to the aspect of informal institutions as substitutes for formal institutions in the following sections.

12 Building on the work of, among others, Heller (1996) and Fox (1996).

13 See, for instance, Rose (1995) on the case of Russia.

14 It is difficult, if not impossible, to talk of causality in these matters and no attempts to determine causality are made in the following.

15 These findings date back to the 1970s, but seem to be equally relevant today.

16 The term 'contract' is to be understood in a broad sense, not necessarily as a legal contract.

17 Note that complementarity here differs from the previous use of the term. We will use the current definition of the term in the remainder of the chapter.

18 Note that we implicitly assume that bonding social capital is simultaneously present when bridging social capital prevails. An example where the opposite may be the case is new urban settings.

19 In fact, Sen (1999) sees freedom as both the means and the end of development.

20 See, for instance, Chapter 8 in Metz et al (2001).

21 This is a traditional neoclassical approach. Resource scarcity and rational behaviour imply that the exchange process will establish the efficient resource allocation. Most top-down models apply this approach. Different approaches are used in bottom-up models on implementation costs. In line with the neoclassical approach, some studies assume that the economy is efficient in the baseline and consequently assess that only a minor overhead cost component is the necessary implementation cost. Other bottom-up studies focusing on technology assessment assume that a specific implementation policy is carried out, where regulations and technical standards support specific technologies.

22 Even in quadrant 1, additional measures can be necessary, as will be illustrated later in the chapter.

23 See for instance, Martinot and McDoom (2000), Chapters 5, 7 and 8 in Metz et al (2001) and Halsnaes, Callaway and Meyer (1998).

24 See, for instance, ALGAS (1999).

25 GHG mitigation policies clearly cannot and should not address all aspects related to deficiencies in the workings of a given economy. Assessing the functioning of the economy, however, provides a basis for evaluating the appropriateness of different mitigation policies and points to areas where minor additional interventions may be associated with benefits.

26 The cost aspects are addressed in more detail later.

27 This is basically a second-best problem related to government policy externalities (or failures) and its relevance, of course, is not limited to this specific quadrant.
28 Note that a first-best policy would require a transfer by lump sum taxation. Even an income-tax based transfer is second best.
29 See, for instance, Narayan (1999).
30 It implies simultaneously a forgone benefit because the lamps would have implied a reduced electricity subsidy payment for the utility if they had been installed by the low-consumption consumers.
31 See Martinot and McDoom (2000).
32 Feldman (1994) introduces these three aspects as key defining dimensions of a market.
33 In this case the social capacity implies the isolation of an option.
34 Social costs = private costs + the costs of externalities.
35 The approach is a further development of a methodological framework which was established as part of the UNEP/GEF project Economics of GHG Limitations (Halsnaes, Callaway and Meyer, 1998; Markandya, 1998). See Appendix 4.1 and Chapter 7 for a more detailed description of the approach, the case studies and the data used.
36 In Chapter 7 two social cost estimates are presented – one including the costs of health impacts in coal-mining and one excluding such impacts.
37 Table 4.6 takes the basic cost data as given in Chapter 7, Table 7.5. In addition, the costs of implementation and the difference in interest rates have been added as described above.

REFERENCES

Alderman, H and Paxson, C H (1994) 'Do the Poor Insure? A Synthesis of the Literature on Risk and Consumption in Developing Countries' in E L Bacha (ed) *Economics in a Changing World: Proceedings of the Tenth World Congress of the International Economic Association,* International Economic Association, London
ALGAS (1999) *Profiles of Investment and Technical Assistance. Greenhouse Gas Abatement Projects,* Asian Development Bank, Global Environment Facility, United Nations Development Programme, Manila, The Philippines
Atkinson, A B (1970) 'On the Measurement of Inequality', *Journal of Economic Theory,* Vol 2, No 3, pp244–263
Bardhan, P K (1981) 'Terms and Conditions of Labor Contracts in Agriculture: Results of a Survey in West Bangal 1979', *Oxford Bulletin of Economics and Statistics,* February
Bardhan, P K (1985) *The State, Society and Economic Growth,* Ananda Publishers, Calcutta
Bardhan, P K (1995) 'The State and Dynamic Comparative Advantage' in K Banerji and T Vakil (eds) *India: Joining the World Economy,* Tata McGraw Hill, New Delhi
Barr, A (1998) *Enterprise Performance and the Functional Diversity of Social Capital,* Centre for the Study of African Economies, Institute of Economics and Statistics, Working Paper Series 98–1, University of Oxford
Barr, A (2000) 'Social Capital and Technical Information Flows in the Ghanaian Manufacturing Sector' *Oxford Economic Papers 52 (539–559),* Oxford University Press
Coleman, J (1990) *Foundations of Social Theory,* The Belknap Press of Harvard University Press, Cambridge, MA
Dasgupta, P (1999) 'Economic Progress and the Idea of Social Capital' in P Dasgupta and I Serageldin (eds) (1999) *Social Capital: A Multifaceted Perspective,* The World Bank, Washington, DC
Dasgupta, P and Serageldin, I (1999) *Social Capital: A Multifaceted Perspective,* The World Bank, Washington, DC
Elliot, P and Booth, R (1993) 'Brazil Biomass Demonstration Project', *Special Project Brief,* Shell International Petroleum Company, London
Ensminger, J (1992) *Making a Market: The Institutional Transformation of an African Society,* Cambridge University Press, Cambridge

Evans, P (1992) 'The State as Problem and Solution: Predation, Embedded Autonomy and Adjustment' in S Haggard and R Kaufman (eds) *The Politics of Economic Adjustment: International Constraints, Distributive Politics, and the State,* Princeton University Press, Princeton, NJ

Evans, P (1995) *Embedded Autonomy: States and Industrial Transformation,* Princeton University Press, Princeton, NJ

Evans, P (1996) 'Government Action, Social Capital and Development: Reviewing the Evidence on Synergy', *World Development,* Vol 24, No 6

Feldman, S (1994) 'Market Transformation: Hot Topic or Hot Air', *Proceedings of the 1994 ACEEE Summer Study on Energy Efficiency in Buildings,* American Council for an Energy-efficient Economy, Washington, DC

Fox, J (1996) 'How Does Society Thicken? The Political Construction of Social Capital in Rural Mexico', *World Development,* Vol 24, No 6

Gittel, R and Vidal, A (1998) *Community Organizing: Building Social Capital as a Development Strategy,* Sage Publications, Newbury Park, California

Halsnaes, K (1996) 'The Economics of Climate Change Mitigation in Developing Countries', *Energy Policy Special Issue,* Vol 24, No 10/11

Halsnaes, K, Callaway, J M and Meyer, H J (1998) *Methodological Guidelines, Economics of Greenhoouse Gas Limitations, Main Reports,* UNEP Collaborating Centre on Energy and Environment, Risø National Laboratory, Denmark

Heller, P (1996) 'Social Capital as a Product of Class Mobilization and State Intervention: Industrial Workers in Kerela, India', *World Development,* Vol 24, No 6

Kozel, V and Parker, B (2000) 'Integrated approaches to poverty assessment in India' in Bamberger, M (ed) *Integrating quantitative and qualitative research in development projects,* The World Bank, Washington, DC

Kranton, R E (1996) 'Reciprocal Exchange: A self-sustaining system', *The American Economic Review,* Vol 86, No 4

Loury, G (1992) 'The economics of discrimination: Getting to the core of the problem', *Harvard Journal for African American Public Policy,* Vol 1

Markandya, A (1998) *The indirect costs and benefits of greenhouse gas limitations,* Economics of Greenhouse Gas Limitations, Handbook Reports, UNEP Collaborating Centre on Energy and Environment, Risø National Laboratory, Denmark

Martinot, E and Borg, N (1998) 'Energy-efficient lighting programs. Experience and lessons from eight countries', *Energy Policy,* Vol 26, No 14, pp1071–1081, Elsevier, UK

Martinot, E and McDoom, O (2000) *Promoting Energy Efficiency and Renewable Energy. GEF Climate Change Projects and Impacts,* Global Environment Facility, Washington, DC

Metz, B, Davidson, O, Swart, R and Pan, J (eds) (2001): *Third Assessment Report of the Intergovernmental Panel on Climate Change,* Cambridge University Press, Cambridge

Narayan, D (1999) *Bonds and Bridges. Social Capital and Poverty,* World Bank Policy Research Working Paper 2167,World Bank, Washington, DC

Narayan, D and Nyamwaya, D (1996) *Learning from the Poor: A Participatory Poverty Assessment in Kenya,* Environment Department Papers, Participation Series 34, World Bank, Washington, DC

Narayan, D and Pritchett, L (1997) *Cents and Sociability: Household Income and Social Capital in Rural Tanzania,* World Bank, Washington, DC

Nketiah, K S, Hagan, E B and Addo, S T (1988) *The Charcoal Cycle in Ghana. A Baseline Study,* Project GHA/82/020, Activity 2.3.5, UNDP/National Energy Board, Ghana

North, D C (1990) *Institutions, Institutional Change, and Economic Performance,* Cambridge University Press, Cambridge

Pamuk, A (2000) 'Informal Institutional Arrangements in Credit, Land Markets and Infrastructure Delivery in Trinidad', *International Journal of Urban and Regional Research,* Vol 24, No 2

Pathan, R, Arul, N and Poffenberger, M (1993) *Forest Protection Committees in Gujarat – Joint Management Initiatives,* Reference Paper 8, prepared for Sustainable Forest Management Conference, sponsored by the Ford Foundation, Delhi

Portes, A (ed) (1995) *The Economic Sociology of Immigration: Essays on Networks, Ethnicity and Entrepreneurship*, Russell Sage Foundation, New York

Portes, A and Landolt, P (1996) 'The Downside of Social Capital', *The American Prospect*, Vol 26

Putnam, R D, Leonardi, R and Nanetti, R Y (1993) *Making Democracy Work: Civic Traditions in Modern Italy*, Princeton University Press, Princeton, NJ

Rose, R (1995) 'Russia as an Hour-glass Society: A Constitution without Citizens', *East European Constitutional Review*, Vol 4, No 3

Rubio, M (1997) 'Perverse Social Capital: Some Evidence from Columbia', *Journal of economic issues*, Vol 31, No 3

Sathaye, J et al (1994) 'Economic Analysis of Ilumex. A project to promote energy-efficient residential lighting in Mexico', *Energy Policy*, Vol 22, No 2

Sen, A (1992) *Inequality Reexamined*, Clarendon Press, Oxford

Sen, A (1999) *Development as Freedom*, Alfred A Knopf, New York

Simmel, G (1971) [1908] 'Group Expansion and the Development of Individuality' in D Levine (ed) *Georg Simmel: In Individuality and Social Forms*, University of Chicago Press, Chicago

Todaro, M P (1997) *Economic Development*, Addison Wesley Longman, London, 6th edn

UNEP (1994) *UNEP Greenhouse Gas Abatement Costing Studies*, Vol 1, 2 and Guidelines Annex, UNEP Collaborating Centre on Energy and Environment, Risø National Laboratory, Denmark

UNDP (1996) *Energy after Rio. Prospects and Challenges*, United Nations Publications Sales, No E.97 III.B.11

Uphoff, N (1992) *Learning from Gal Oya: Possibilities for Participatory Development and Post-Newtonian Social Science*, Cornell University Press, Ithaca, New York

Woolcock, M (1998) 'Social Capital and Economic Development: Toward a Theoretical Synthesis and Policy Framework', *Theory and Society*, Vol 27, pp151–208, Kluwer Academic Publishers, Kingston-upon-Thames

Woolcock, M and Narayan, D (2000) 'Social Capital: Implications for Development Theory, Research, and Policy', *The World Bank Observer*, Vol 15, No 2

World Bank (1996) *Rural Energy and Development: Improving Energy Supplies for 2 Billion People*, Washington, DC

World Bank (1997a) *The Greening of Economic Policy Reform*, Vol 1, World Bank, Washington, DC

World Bank (1997b) *Expanding the Measure of Wealth: Indicators of Environmentally Sustainable Development*, World Bank, Washington, DC

Young, O R (ed) *Institutional Dimensions of Global Environmental Change*, IDGEC Science Plan, IHDP, Bonn

APPENDIX 4.1: OVERVIEW OF THE COST CALCULATIONS

The approach is based on Markandya (1998).

Employment Benefits

To assess the social benefits of employment, the following factors need to be taken into account: the unemployment rate, the period of employment, the extent of any unemployment benefit and the health benefits of employment.

The road pavement project has social employment benefits that are estimated in the following way. The project implies new employment for unskilled workers in the construction phase of 1858 man-years. Based on Botswana data, their annual wage is US$1200. The lost value of non-work time is set as 15 per cent of the wages, which equals $180 per man-year, the unemployment rate is 25 per cent and the unemployment benefit is $0 per year.

The health benefits of employment per person per year are estimated at:

6/1000 * 0.75 * 1085 = US$4882.5

where:

US$1085 is the value of statistical life for Botswana for an income elasticity of one;
0.75 is the excess mortality rate for unemployed persons; and
6/1000 is the assumed age specific death rate in Botswana for males.

The total social benefits of employment are then:

a   Health benefit: US$4882.5*0.25*1858 = US$2,267,921
b   Lost value of non-work time: US$180*0.25*1858 = –US$83,610
c   Employment benefit: US$1200*0.25*1858 = US$557,400
     Total benefits: a+b+c = $2,741,711

These benefits have to be deducted from the actual labour costs of the project.

Time Savings

In the road pavement project, the benefits of reduced time spent on transportation will fall to the users of the paved road and are estimated on the basis of the following data:

Total number of vehicles travelling the route: 64,605
Distance travelled, vehicles-km/year: 18,409,505

The value of time is estimated on the basis of the average annual GDP per capita, which is US$7020 (PPP, current 1995 prices). Assuming a 45 hours working week, gives an estimated value of time of (7020/52)/ 45 = US$3 per hour.

Time difference: based on the project information, it is assumed that the average speed will be 100km/hour instead of 40km/hour. This means that the time saved on the entire stretch of 531km is 7.965 hours or 0.9 minutes per km.

Total time savings: 0.9*18,409,505/60 = 27,6142.58 hours saved
Total value of time savings: US$3 * 27,6142.58 = US$828,428

Income Distribution Benefits

As we saw in Chapter 2, a key issue in evaluating climate change policies is their impact on intragenerational equity, in which one impact indicator is the income distributional consequences of the policies seen in a national context. Following Markandya (1998), it is possible to use 'income weights', so that impacts on individuals with low incomes are given greater weight than those on individuals with high incomes.[1]

The costs of different GHG programmes, as well as any related benefits, belong to individuals from different income classes. Economic cost-benefit analysis has developed a method of weighting the benefits and costs according to who is impacted. This is based on converting changes in income into changes in welfare, and assumes that an addition to the welfare of those on a lower income is worth more than an addition of welfare to richer people. More specifically, a special form can be taken for the social welfare function, and a common one that has been adopted is that of Atkinson (1970). He assumes that social welfare is given by the function:

$$W = \sum_{i=1}^{N} \frac{A Y_i^{1-\varepsilon}}{1-\varepsilon}$$

where:

W is the social welfare function,
Y_i is the income of individual i,
ε is the elasticity of social marginal utility of income or inequality aversion parameter, and
A is a constant.

The social marginal utility of income is defined as:

$$\frac{\partial W}{\partial Y_i} = A Y_I^{-\varepsilon}$$

Taking per capita national income, $\bar{Y}$, as the numeraire, and giving it a value of one gives:

$$\frac{\partial W}{\partial Y_i} = A \bar{Y}^{-\varepsilon} = 1$$

and

$$\frac{\partial W / \partial Y_I}{\partial W / \partial \bar{Y}} = \left[\frac{\bar{Y}}{Y_I} \right]^{\varepsilon}$$

In this way the marginal social welfare impact of income changes by individuals is the elasticity of the ratio of the per capita income $\bar{Y}$ and the income of individual i, Y_i. The marginal social welfare impact of income changes by individual i also can be denoted as SMU_i, where SMU_i is the social marginal utility of a small amount of income going to individual i relative to income going to a person with the average per capita income. The values of SMU_i are, in fact, the weights to be attached to costs and benefits to groups relative to different cost and benefit components.

To apply the method, estimates of $\bar{Y}$ and ε are required. Markandya (1998) reports estimates of the inequality aversion parameter (ε) in the range 1–2.

In this chapter an inequality aversion parameter of 1 is applied, implying that environmental damages are valued to all individuals at the value associated with the average income individual and that the 'income elasticity' of environmental damage with respect to income is assumed to be one.

With respect to the Botswana examples, low-income persons will be the beneficiaries of employment creation in the road pavement example. Following the above, the value of these transfers should be increased according to the ratio of the average income of the beneficiaries relative to the average. We assume that the beneficiaries have an income of 30 per cent of the average, leading to an adjustment coefficient of 3.1^2. Adjusting the employment benefits by this factor gives the following benefit:

Employment benefit (US$000): 8499.3

In the road pavement project, it is assumed that the users are at the average level of income, hence no income distribution adjustment is made for the time-saving benefits.

Based on employment data from Botswana, it is also assumed that the health benefits of decreased coal production in the mines accrue to people at the average level of income, which means that these benefits have not been adjusted for income distribution effects.

Direct Cost Savings due to Learning Effects

To include the learning effects, it is assumed that implementation is associated with costs that amount to US$7222.5, which is 15 per cent of total capital and operation and maintenance costs. These costs are added in the gross financial cost assessment. To reflect learning effects, these costs are assumed to decrease over the project duration and this is reflected in the net financial costs.

The same type of adjustment can be made to reflect indirect cost saving due to learning.

Market Development Effects

To provide an example of the effects of local business and market development, we split the efficient industrial boiler example into two cases. In the first case local market and business development is low, which is reflected in high interest rates due to the lack of credit opportunities. We have applied a discount rate of 13 per cent in case 1 to mirror this.

In the second case we assume that another project on cash crop production and marketing has been implemented. This project is totally unrelated to the GHG emission-reduction option but is associated with positive externalities, as local markets for credit have been developed as a consequence of the project. A discount rate of 10 per cent has been applied in case 2 to reflect this effect.

NOTES

1   It should be noted that while a number of analysts do not support the use of such weights, some do and policy-makers sometimes find an assessment that uses income weights useful.
2   This coefficient is taken from Markandya (1998, p35). Note that the average income of the beneficiaries may well be less than 30 per cent of the average income

Chapter 5

Analytical Approaches for Decision-making, Sustainable Development and Greenhouse Gas Emission-reduction Policies

Kirsten Halsnaes and Anil Markandya

INTRODUCTION AND SCOPE

So far we have discussed the issues that arise with the introduction of sustainable development into the greenhouse gas (GHG) emission-reduction policy. Chapter 2 outlined the theoretical issues from a sustainable development perspective, Chapter 3 provided a summary of the discussions in the climate change literature and Chapter 4 discussed the concept of social capital in some detail. In this chapter we focus on decision-making rules for ranking GHG projects.

The chapter outlines three approaches for how GHG emission-reduction policies can be assessed in the context of sustainable development – ie, including the economic, environmental and social dimensions of this concept. The idea is to outline a methodological framework for interactive policy decisions and technical assessments that is based on a number of distinct analytical steps.

The main differences and similarities in alternative approaches that can be used to assess sustainable development dimensions are highlighted in order to explain why specific study assumptions and analytical structures lead to different policy conclusions. The sustainable development dimensions can be assessed with a number of different analytical approaches, and the chapter includes a comparable assessment of the ones most commonly used in relation to studies of GHG emission-reduction policies in developing countries. The approaches presented are cost-benefit analysis (CBA), cost-effectiveness analysis (CEA), and multi-criteria analysis (MCA).

A number of case examples of GHG emission-reduction projects are reviewed in the chapter using these three alternative technical approaches. This leads to a discussion about the wide range of alternative policy recommendations that can be generated on the basis of different assumptions and analytical structures.

DECISION-MAKING APPROACHES AND SUSTAINABLE DEVELOPMENT CONCEPTS

As we noted in Chapter 2, sustainable development can be defined in many ways – the literature includes hundreds of alternative definitions (Pezzey, 1992; Pearce, Barbier and Markandya, 1990). A commonality between the various definitions of sustainable development, however, is that they try to integrate a broad range of developmental, environmental and social dimensions in a short- and long-term timeframe. Consequently, sustainable development evaluations by their basic nature involve an evaluation of multiple policy objectives. There is therefore a need to support sustainable development policy evaluations with technical approaches that can facilitate a systematic assessment of multiple objectives, including considerations of trade-offs between different objectives and constraints to reflect non-substitutable resources.

The assessment of multiple policy objectives can draw on a number of experiences from the literature on public planning and decision-making (Roseland, 2000; Johansen, 1979; Bogetoft and Pruzan, 1997). The application of public planning approaches to the evaluation of sustainable development dimensions is understood by Roseland (2000) as an activity where planning is a decision process that includes several stages, beginning with the identification of goals that will structure the decision[1] and ending with programme analysis, where the impacts of different programme alternatives are evaluated. Such an approach can be understood as a rational planning model, in which a technical assessment is used to consider the impacts of given options on specific policy priorities. A rational planning model first of all can be helpful in the establishment of systematic quantitative and qualitative information related to a chosen set of sustainable development indicators that are considered to be important in relation to specific policy priorities. A number of sustainable development aspects, however, are not very well represented by indicators. Examples of such factors are social and institutional capacity-building, which cannot be turned easily into an indicator that is 'measured' in a rational planning model (see Chapter 4 on social capital, which includes an extensive discussion about potential indicators and measurement standards).

The idea of the methodological approach outlined in this chapter is to transform general sustainable development policy priorities into indicators that can be used for decision-making, to select between specific policy options that reduce GHG emissions. In this there will be a tendency to pay most attention to those sustainable development policy dimensions that can be quantified relatively easily. This means, for example, that economic impacts, material resource consumption, and impacts on the physical environment tend to be included. More limited attention, however, will be given to a number of social sustainability dimensions and to longer-term impacts, such as intergenerational considerations, which can be fully addressed only on the basis of a broader evaluation, as discussed in Chapter 2.

A number of complexities arise in the assessment of projects that impact on several sustainable development policy objectives and it is difficult to define a consistent framework that reflects the different scope and character of the joint impacts. Furthermore, the inclusion of GHG emission reductions as an objective in a broader policy agenda imposes a number of additional evaluation problems due to the long-term and very uncertain character of the climate change problem (Banuri et al, 2001). Climate change is a global environmental problem that originates from the accumulated stock of atmospheric greenhouse gases that has been built up over the last hundred years. Damages in the form of climate change are very uncertain and there is no direct relationship between the GHG emission from individual countries and the climate change damages that this country will experience.

It is also evident that the values attached to the different outcomes of a GHG policy will vary, according to whether the policy-maker represents the interests of the developing country, the industrialized country or an agency responsible for providing global assistance for GHG projects. Some policies, for example, seem to be attractive to international donors representing industrialized countries due to their capacity for reducing GHG emissions, while other policies may have smaller GHG emission-reduction impacts but a number of joint policy impacts that are attractive when seen from a local development perspective. The final selection of policy options and the design of financial transfer schemes to support the implementation will be a matter of negotiation therefore, given the different perspectives of the parties.

LINKAGES BETWEEN BROADER POLICY PRIORITIES AND TECHNICAL ASSESSMENTS

Before a formal technical assessment can be undertaken, it is essential to establish what are the policy priorities. Inevitably this has a normative aspect to it. In order to reflect this normative element it is recommended to include a separate analytical step that addresses linkages between general policy priorities and the recognition of these in the objective function of the technical assessment.[2] The advantages of such an approach are that the normative aspects implicit in social welfare functions and other sorts of objective functions are considered explicitly. A way to structure the interaction between broader policy evaluations and the technical assessment is to go through a number of separate steps like the ones suggested in the following list.

Policy evaluation steps:

- The selection of policy priorities that are expected to be relevant to the planning problem under consideration. These priorities can be related, for example, to political decisions or to official plans that have been developed in other policy contexts.
- Considerations on the relative value or priority of different policy impact areas.
- Initial screening of projects that are considered to be relevant and that should be included in the assessment.
- Final discussion and conclusions based on the output from the technical assessment.

Technical assessment steps:

- Definition of indicators that reflects the various policy objectives introduced in the policy evaluation step. This involves the definition of measurement standards for the indicators and aggregation rules.
- General procedures and rules that can be used for an integrated assessment of various indicators. This includes an outline of an approach for the handling of trade-offs between different objectives.
- The preparation of technical output that can be supplied to policy-makers.

An example of the linkages between general development policy objectives and the formal application of these into the objective function of a technical analysis is provided by the following example of a GHG emission-reduction project. Suppose the reference energy system is a coal-based power system using imported coal and that other major energy sector activities are woodfuel and charcoal consumption for cooking, and petroleum and coal consumption for industrial boilers. A number of national development programmes and sectoral plans have given high priority to economic growth, in particular in the manufacturing sector, and to employment generation, local air pollution control, rural development and poverty alleviation. All these development goals are considered to be closely related to the potential impacts of implementing the GHG emission-reduction projects considered. Table 5.1 below illustrates how such general development priorities can be specific as indicators that are included in the objective function of a technical assessment.

Table 5.1 *Examples of linkages between general policy priorities and indicators that can be integrated in technical assessments*

Policy priorities in general development programme	Examples of arguments included in the technical assessment
Economic growth	Macroeconomic indicators – eg GDP growth
	Social cost of the project
Employment	Impacts on employment for different labour market segments
Rural development programmes	Economic activity generated in rural area
	Energy supply to rural area
Local air pollution improvement	SO_2, NO_x and particulate emissions
	Acid depositions
	Health impacts
Increasing activity in the manufacturing sector	Investments in manufacture
	Energy supply to manufacture
Poverty alleviation	Change in numbers under various poverty lines

The examples of indicators that are included in Table 5.1 are just potential applications chosen out of a range of options and the selection of the actual indicators that are evaluated in the technical analysis in this way has a normative character.

Chapter 7, which provides detailed case studies for GHG emission-reduction projects in Zimbabwe, Botswana, Mauritius and Thailand, includes a set of specifically chosen indicators similar to those in Table 5.1. The indicators for these countries are financial and social costs, local air pollution, income distribution, health impacts and employment generation.

ANALYTICAL APPROACHES

The focus of this section is to make a comparable assessment of alternative analytical approaches that can be used for an integrated assessment of GHG emission reductions and sustainable development impacts. A large number of different approaches in principle can be used for such an evaluation, including general modelling and scenario approaches, sectoral and project evaluation approaches, and various decision support tools that establish a dialogue between broader policy processes and expert judgements (Markandya et al, 2001; Toth et al, 2001). Some of the approaches that most commonly have been used to evaluate GHG emission-reduction policy options in studies for developing countries are: CBA (Squire and van der Tak, 1975; Ray, 1984), CEA (Markandya, Halsnaes and Milborrow, 1998; Sathaye, Norgaard and Makundi, 1993), and MCA (Keeney and Raiffa, 1993). Given the prominence these received, the following discussion focuses on them and considers how they can be developed to facilitate the integration of broader development objectives.[3]

Before going into the three approaches in detail, we should note that they all assume the policy priorities to be given exogenously, perhaps determined by the political process.[4] There are other approaches which do not make this assumption (Jasanoff and Wynne, 1998). In these, policy-makers and technical experts undertake a common planning process that discusses policy priorities and transforms these to specific objectives, and approaches that separate policy making and technical assessments through the use of stated policy preferences as exogenous input to technical assessments. Stated policy preferences can be reflected, for example, in national development and sectoral programmes (Cantor and Yohe, 1998; Bogetoft and Pruzan, 1997).

In the remainder of this chapter a structure for the evaluation of various policy impacts is outlined, and screening rules and procedures for the evaluation of trade-offs between different impacts are suggested. Similarities and differences between the structure and decision rules of CBA, CEA and MCA are illustrated in relation to the different subcomponents of the policy impact evaluation.

Formalization of the Decision-making Problem

The CBA, CEA, and MCA approaches are considered in this section as general categories of decision support tools, and a common structure for using these to evaluate specific policy options (which in this context take the starting point in options that reduce GHG emissions) in relation to multiple policy objectives is outlined below.

The technical assessment of GHG emission-reduction options in the context of multiple policy priorities can be structured around the following analytical steps:

1   Identification of a set of M possible GHG emission-reduction options. Call this set $GHGP$ and call any one option $GHGP_i$
2   Selection of a set of state variables (X) considered to represent areas that will be influenced by project implementation. Assume that X is a N dimensional vector – ie, there are N variables that measure the impacts at any point in time $(X = [x_1, x_2 \ldots x_n])$. The state variables x are dependent on the structure and performance of the economy and are reflecting national policies as well as exogenous factors.
3   Description of the non-policy case (baseline case) with regard to the chosen set of state variables x, defined for a given time horizon T. Call this set X_0, where X_0 is defined over the N. $X_0 = (x_{10t}, x_{20t} \ldots x_{n0t})$, t= 1, 2, ... T.
4   Assessment of the performance of the economy over the state variables, given the implementation of the GHG emission-reduction policy options, X_i. Call this assessment A_i. A_i will depend, of course, on the values of X for the policy options.
5   Specification of the national preference scale for sustainable development policies with regard to the state variables of the economy X represented by $W(X)$. $W(X)$ is a function that orders the preferences regarding to X.
6   Assessment of $W(X_i)$ and $W(X_0)$.
7   Evaluation of $(W(X_i) – W(X_0))$.

Repeat this for each option.

This formalized structure for integrating national preferences for sustainable development policies into technical assessments in summary can be said to include the following four elements:

1   **Project screening** The screening of projects and definition of the set of possible policy options.
2   **Definition of system boundary** The selection of state variables x, that reflects the focal policy areas where the policy options are expected to have major impacts.
3   **Definition of a standard for measuring policy impacts** The formalization of a procedure for assessing the state of the economy x. The assessment will consider the non-policy case X_0 and the policy case X_i.
4   **Valuation of policy impacts** The selection and formalization of national preferences for sustainable development policy objectives that are represented by $W(X)$.

The sustainable development evaluation can also be illustrated in Figure 5.1 below, where X_0 describes the state of the economy before the implementation of a project and X_i the state of the economy with regards to the same state variables after implementation of the policy option. The function $W(x)$ represents the evaluation of the change from the state X_0 to X_1.

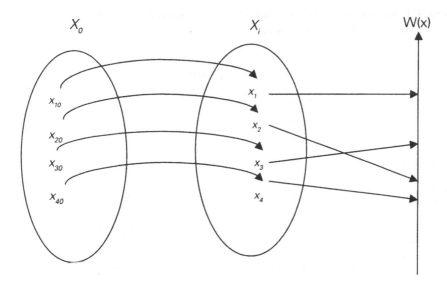

Figure 5.1 *Structure for the evaluation of transformations from a state of the economy X_0) to the state X_i, given the preference function $W(X)$*

The structure of the technical assessments can also be illustrated in relation to two examples of national GHG emission-reduction projects. These are a rural biogas plant for power generation based on waste and a wind turbine for power generation, and GHGP is here constituted by these two projects. The state of the energy supply and economic development in the rural area affected is described by X_{it}, the individual vector components of which include data on energy consumption for end uses such as cooking, lighting and water pumping, air pollution impacts, local employment and project costs.

The implication of implementing projects belonging to GHGP on each subcomponent in X_i is assessed. For each subcomponent in X_i a quantitative or qualitative 'score' is now given to the outcome on the components as illustrated in Table 5.2. Note that the table presents only one time period. Usually each project will have values of the variables over several periods.

Returning to the previously listed four formalized assessment steps, the matrix shown in Table 5.1 covers step 1 (project screening), step 2 (system boundary) and step 3 (standard for measuring policy impacts). All together these steps involve a technical specification of policy variables and impacts, and the definition of a standard for measuring project impacts, and the information generated at each step in these steps leads up to the final evaluation of policy impacts as the last stage of the assessment. This final stage is the most complicated and controversial part of the evaluation exercise because it involves the definition of a standard for cross-cutting the comparison of the individual X_i components.

An evaluation of multidimensional policy impacts like the outcomes on the x_i's involves the comparison of impact areas that by nature is somehow incommensurable. Even more than the two project cases given above, the policy impacts can include various environmental areas, and social and economic

Table 5.2 *Matrix for evaluating impacts of case examples in the form of a biogas project and a wind-turbine project*

	Non-policy state	State with implementation of biogas project	State with implementation of wind-turbine project	Impacts of implementing biogas project	Impacts of implementing wind-turbine project
Cost	x_{10}	BX_1	TX_1	$(BX_1 - x_{10})$	$(TX_1 - x_{10})$
Energy consumption	x_{20}	BX_2	TX_2	$(BX_2 - x_{20})$	$(TX_2 - x_{20})$
Local environment	x_{30}	BX_3	TX_3	$(BX_3 - x_{30})$	$(TX_3 - x_{30})$
Employment	x_{40}	BX_4	TX_4	$(BX_4 - x_{40})$	$(TX_4 - x_{40})$
GHG emissions	x_{50}	BX_5	TX_5	$(BX_5 - x_{50})$	$(TX_1 - x_{50})$

dimensions that are difficult to measure in common units. Furthermore some of the impacts are characterised by major uncertainties and include elements of non-substitutable values and irreversibility (Markandya et al, 2001). These methodological problems constitute a major drawback in the application of general valuation rules to the assessment of multiple policy impacts, and it is therefore desirable, as far as possible, to limit the formal evaluation of trade-offs between different policy impacts that appear in areas outside the scope of traditional economic evaluations.[5]

The following section outlines a systematic approach for narrowing down the number cases where trade-offs are considered as part of a policy evaluation.

Screening Rules for Policy Evaluation

The evaluation of policy options as previously argued can be understood as a sort of a planning exercise where an estimate is made of how given policy priorities can be met with the lowest possible resource requirements. This implies that policy options that offer a larger output on prioritized impact areas, given constant resource requirements, are preferred to options that offer lower output. This understanding of the planning problem suggests the inclusion of an evaluation of the relative efficiency of policy options as part of a policy screening procedure. Such a mechanism is illustrated in Figure 5.2.

The idea of Figure 5.2 is to structure the assessment of policy options into four distinct quadrants, Q1 to Q4, each representing different subsets of potential trade-offs between the impacts of a given set of policy options represented by options B, C, D and E relative to the policy option A. For simplicity, the policy impacts are in this context transformed into a two-dimensional impact vector.

The performance of the policy options compared with A differs in the four quadrants Q1, Q2, Q3 and Q4 that is drawn with A as the centre. All policy options located in Q1, such as option D, have a larger impact on policy impacts 1 and 2 than A, and are in this way considered to be more efficient than A. The

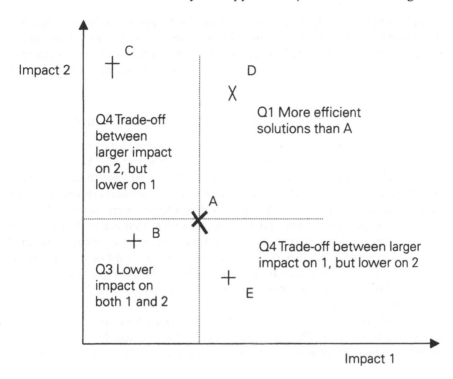

Figure 5.2 *Illustration of an approach for analysing the efficiency and trade-offs of policy impacts measured in comparison with the performance of a project A*

opposite is the case for options located in Q3, like option B, where the impacts on both objective 1 and 2 are smaller than for A – policy options located in Q3 then can be said to be less efficient than option A. A trade-off between a larger impact on policy objective 1, but a smaller impact on 2 is seen for policy options located in Q2, like option C, and the opposite is the case for options in Q4, like option E, where the impact is larger on policy objective 2, but smaller on objective 1 compared with project A. The efficiency criteria applied to the evaluation of options relative to A then will imply that option D is more efficient than A and option B is less efficient than A, and this conclusion can be drawn without considering trade-offs between impact 1 and 2. The assessment of projects like E and C in relation to A, however, requires the use of a formalized approach for evaluating trade-offs.

The evaluation of trade-offs between different policy impacts can also be limited to fewer policy options by the introduction of constraints for some of the impacts considered. Such constraints can reflect, for example, safe minimum standards for specific ecological systems. The use of safe minimum standards to reflect the non-substitutable character of, for example, natural capital as part of a sustainable development evaluation has been suggested in a number of sustainability concepts including, for example, the strong sustainability introduced by Daly (1990).

Safe minimum standard decision rules can be defined for individual projects or for a portfolio of projects. This principle is illustrated in Figure 5.3 applied to the evaluation of the projects A, B, C and D. A constraint, cx_2, is in this case put on impact 2, and the projects that meet this constraint are now limited to be projects C and D.

Evaluation of Trade-offs Between Multiple Policy Impacts

The following section provides a brief overview of the measurement and valuation approaches that typically are used in CBA, CEA and MCA.

The evaluation of trade-offs requires the use of a preference function that includes information about the relative value of the individual arguments in the objective function as described by the individual vector components of X_i. This preference function can be defined as a function $W(x)$ that is a preference scale for development policies regarding X. A preferable strategy then can be defined in various ways, depending on 'rules' for substitution of the different dimensions of the development preferences. In this way $W(X)$ can be used to reflect an evaluation of sustainable development policy objectives that are based on indicators represented by the x_i's.

A simplified operational approach for a preference function on sustainable development follows. Let the preference function for sustainable development be denoted W, where W is a single valued function defined of the set of state variables X.

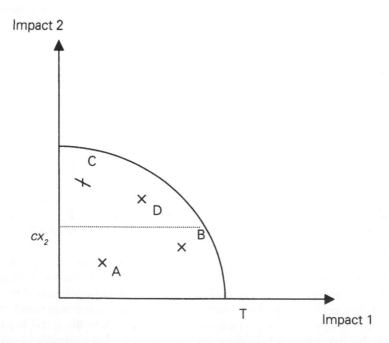

Figure 5.3 *Structure for the evaluation of projects with regard to policy impacts 1 and 2, given a minimum constraint on impact 2 of* cx_2

The preference ordering is assumed to have the properties usually associated with such orderings. These are:[6]

Completeness: for any feasible policy options X_i and X_j, either $X_i \to X_j$ or $X_j \to X_i$ or $X_i \approx X_j$. We use $\to$ to stand for 'is preferred to' and $\approx$ to stand for 'is indifferent to'.

Transitivity consider three policy options X_i, X_j, and X_k. If $X_i \to X_j$ and $X_j \to X_k$, then it must be the case that $X_i \to X_k$.

If these assumptions hold, a single valued continuous preference function W can be shown to exist and to have the property:

$W(X_i) > W(X_j)$ if and only if $X_i \to X_j$
$W(X_i) = W(X_j)$ if and only if $X_i \approx X_j$

The specification of W reflects different sustainability preferences and, of course, presuming its existence is a major assumption. The most important implication is that the different components of X can be measured.

The above is an abstract and general representation of preference functions. We now consider specific representations of W.

Preference Function Using an Extended CBA

One of the most commonly used preference functions in CBA is a social welfare function that is constructed as the aggregate of the welfare of all individuals. This can be represented as follows.

Let the function W be defined as:

$$W^i = \sum_j V^j(x_{10}^j, x_{20}^j, x_{30}^j...x_{N0}^j) \qquad (5.1)$$

Where, as before, x_{k0}^j is the level of impact k, under the baseline on individual or agent j. If the option under consideration changes the values of the x's, we can, under certain conditions, write the change as:

$$DW^i = \sum_j \sum_{i=1}^{i=N} \frac{\partial V^j}{\partial x_i^j} dx_i^j \qquad (5.2)$$

But $\dfrac{\partial V^j}{\partial x_i^j}$ can be interpreted as the marginal willingness to pay (MWTP) by individual or agent j for the change in impact i. Under competitive conditions and if the 'impact' i can be interpreted as a commodity, it will have a price, p_i. If the impact does not have a price it is still meaningful to refer to the term as the MWTP. Suppose for simplicity that the impacts can be partitioned into two groups, $1 \ldots L$ which have market prices and $L+1 \ldots n$ which do not. Then equation 5.2 can be rewritten as:

$$DW^i = \sum_{i=1}^{i=L} p_i \sum_j dx_i^j + \sum_j \sum_{i=L+1}^{i=N} MWTP_{ij} dx_i^j \qquad (5.3)$$

The first term is simply the change in the market value of the commodities x_i. The second term is the value of the impacts expressed in terms of the MWTP. These have to be assessed by indirect methods as such preferences are not directly represented in market data. In Chapter 7 we discuss some of the methods by which such estimates could be obtained. If the data can be elicited, then equation 5.3 provides one-way concrete representation of the function W.[7]

Preference Function Applied to Cost-effectiveness Analysis

In the case of CEA, the situation is somewhat more complicated. The tool is particularly useful when there are a number of options, such as those discussed above, and the following other conditions hold:

(a)   there is a unique product or target that the options seek to achieve – for example, a reduction in GHGs emitted per annum relative to a defined baseline; and

(b)   a fixed amount of reduction that can be achieved from each option.

In that case, the following welfare function can be defined:

$$W = \sum_i (v - C_i)\Delta Z_i \tag{5.4}$$

Where v is the unit value of the target/output Z, C_i is the cost per unit of reduction in the Z and ΔZ_i is the change in the target achieved from option i. The cost C includes both direct costs as well as those costs that cannot be measured using market data (eg environmental and employment impacts).

It can be shown that for such a welfare function the following holds.

If W is maximized subject to a constraint on the total amount of change in Z and if there are limits on the amount of ΔZ_i, then at the optimum it will never be possible for a project with a higher unit cost to be undertaken before one with a lower unit cost has been exhausted. In other words, the policy-maker should always select the lowest cost options first.

Given that much of the GHG policy is concerned with precisely this kind of situation, it is not surprising that the CEA rule has proved popular. However, it is also important to note that, in the absence of the above conditions, the use of a CEA rule does not guarantee that the benefits of a GHG mitigation programme will be maximized. This can be shown with a simple example:

Suppose we have two options: A and B
The cost per unit reduction from option A is US$5/tonne
The cost per unit reduction from option B is US$7/tonne
The value of a unit reduction in both cases is US$20/tonne
Option A can achieve a reduction of 100 tonnes and option B can achieve a reduction of 500 tonnes

Then the net benefit from A is: 100 x (20 – 5) = US$1500
And the net benefit from B is: 500 x (20 – 7) = US$6500

The reason is that CEA does not allow for the scale effect. Of course, if there is a target reduction of, say, 500 units to be made then it is best to do all of A and 400 units of B, which has a total benefit of US$6700. But if the options are mutually exclusive or there is no limit on how many can be carried out, then B comes before A.

This may be important for a debate about the potential use of the Clean Development Mechanism (CDM) defined by the Kyoto Protocol (UNFCCC, 1997) to facilitate that countries that are committed to reduce its GHG emissions according to the protocol can earn credits from financing emission-reduction projects in developing countries. As we know, developing countries have no targets and therefore do not need to follow the CEA rule in terms of their priorities or in terms of what they offer. It is better that they use the net benefit rule given in the previous section. As far as we are aware, this has not featured in the discussion on the appropriate decision-making rules.

Preference Function Applied to MCA

MCA is a broad category of different decision support systems and the literature includes a wide range of different preference functions. This includes approaches that develop policy priority weights based on suggestions by central decision-makers and approaches that use broader policy evaluation groups to discuss potential weights and to consider trade-offs revealed in the analytical process (Keeney and Raiffa, 1993; Bogetoft and Pruzan, 1997). MCA is also sometimes used to produce a multidimensional set of policy impact information that are not related to integrated decision rules in the form of preference functions.

An efficient policy alternative can be considered as optimal in a rational planning model if the decision-maker's desires can be described by a preference function:

$$W_M = \sum_{i=1}^{n} w_i V_i(x_i) \qquad (5.5)$$

where w_i is policy impact weights and $V(x_i)$ is the value function of x that is a linear in the individual criteria.

The procedure that is to be undertaken to such an optimal solution can be described as a two-step procedure, where the first step is to search for efficient solutions and the second is to involve the decision-maker in choosing a preferred solution. This approach can be illustrated as in Figure 5.4, where the set of feasible alternatives Y is delineated by OAB. The straight lines are iso-value curves depicting (y_1, y_2) combinations that give the same values of the criteria $a_1 y_1 + a_2 y_2$. The iso-value curves in Figure 5.4 illustrates the case where the weights are equal, $a_1 = a_2$. The most efficient solution is in this case Y^* that touches the highest iso-value curve. By varying the importance weights $a_1 \ldots a_n$ it is possible to generate different efficient alternatives. An often applied approach is to determine the weights with a normalisation procedure, where the x_i's for all projects considered are normalized on a scale from zero to one. To make this linear weighting procedure to work in general requires specific assumptions about the feasible alternatives

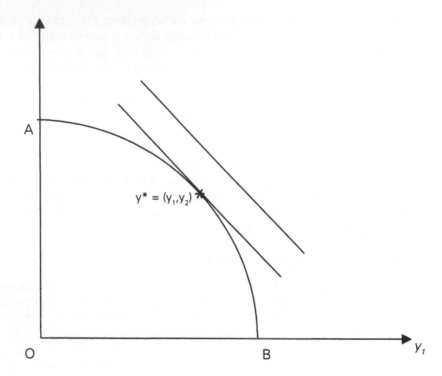

Figure 5.4 *Illustration of a weighting approach in MCA*

Y, which should be convex as a condition for finding efficient solutions to the optimization problem.

It is important to recognize that the use of a normalization procedure for establishing values is very different for the valuation approach used in a welfare function as part of a CBA. The values used in the welfare-based analysis are based on revealed individual preferences, as stated in a broader market context, and reflect in this way the relative value of all goods and services in the economy, while the approach of the MCA as specified above is to determine a relative valuation for the x_i's, given the specific projects considered.

Overview of the Main Structural Differences in the Technical Approaches

The previous sections have identified a number of differences in the preference functions applied to CBA, CEA and MCA, which in particular shows up in relation to the valuation rules that are used to evaluate trade-offs between different policy impacts. In a simple representation of the preference function of the different approaches, one of the major differences that was found between the welfare economic-based CBA and CEA approaches, and the MCA approach, respectively, was that the two first approaches aimed at establishing values of the policy impacts on the basis of market prices or opportunity costs, while the MCA approach aimed at the establishment of a relative ranking of policy impacts among the projects considered. This ranking could be established, for example, in a dialogue with different policy-makers. Some applications of MCA also try to avoid the use of policy weights or valuation rules by inviting a broader audience to an interactive dialogue about the evaluation of trade-offs between different impacts. A systematic overview of some of the main differences in the structure of CBA, CEA and MCA is presented in Table 5.3.

EVALUATION OF CASE PROJECTS WITH DIFFERENT ANALYTICAL APPROACHES

A number of differences between the CBA, the CEA and the MCA are shown in the following section through the evaluation of five case examples of GHG emission-reduction policy options. The different approaches are used to evaluate similar projects and quantitative impact data, and it is therefore possible on this basis to draw a number of conclusions about the implications of the specific weighting rules and decision criteria on policy priority setting.

Table 5.3 *Overview of main structural elements in CBA, CEA and MCA*

	Cost-benefit analysis	Cost-effectiveness analysis	Multicriteria analysis
Selection of state variables x	Based on welfare concepts – eg defined to reflect policy priorities	Partly based on welfare concepts – eg defined to reflect policy priorities	Indicators representing policy priorities
Standard for measurement of x	Welfare, eventually in monetary units	Cost minimization, eventually in monetary units GHG emissions in physical units or other policy goals	Quantitative and/or qualitative units
Weighting rules	Individual preferences as stated on markets	Individual preferences as stated on markets	Alternatives: No weighting Preferences of policy makers Broader policy process
Preference function	Maximize welfare	Minimize costs of achieving a target reduction of GHG	Total score on indicators if weighting rules are applied Individual indicators Sensitivity analysis Trade-off analysis

The five GHG emission-reduction policy options considered in this section have been chosen out of a broader range of options that were included in a climate change mitigation costing study for Botswana. The performance of the five case examples are assessed in relation to the impacts on costs, local air pollution control, employment generation and GHG emissions reduction. The case examples have been assessed with two alternative applications of, respectively, a CBA-based approach, a CEA, and with an MCA-based approach. The comparative assessment leads up to a discussion about differences and similarities between the approaches that focus on the implications of the applied objective functions and the weights assigned to individual impacts.

Introduction of the Case Examples

The Botswana mitigation cost study included an assessment of the direct cost of implementing 20 alternative GHG emission-reduction options in the energy and transport sector. A number of these policy options were included in a more detailed cost-effectiveness analysis of the indirect costs and benefits of GHG emission-reduction policies that included an assessment of financial and social costs, local air pollution impacts in the form of SO_2 and NO_x emission reductions, employment generation, income distribution, and health impacts – the results of that analysis study are reported in Chapter 7.[8] The current section will compare the CEA for five of these case projects with two alternative specifications of a CBA and an MCA approach in order to illustrate the differences and similarities between the approaches.

The Botswana case examples considered in this comparative assessment are:

1   an efficient boiler project for industry;
2   a transportation project where sandy roads are paved;
3   an efficient lighting programme for households and the service sector;
4   a central photovoltaic (PV) plant; and
5   a power factor correction project.

Table 5.4 gives an overview of the projects.

The project evaluation includes the assessment of a number of different cost concepts reflecting financial and social perspectives. The financial cost assessment includes two concepts – namely, gross and net financial costs. Gross financial costs are the costs of running the GHG emission-reduction project, excluding the capital costs of the substituted baseline activity. Net financial costs are the gross financial cost minus the savings in operation and maintenance cost. The difference between the net and gross financial cost concepts can be illustrated in the case of an efficient industrial boiler as a GHG emission-reduction project. The gross financial costs of the project will be the capital cost of the motor, the cost of operation and the maintenance including power, and implementation costs. The net financial costs for this project will be the gross financial costs minus any difference in operation and maintenance cost, and the the value of electricity savings assessed in relation to the baseline case.

The social cost by definition is the sum of the private costs that occur to individuals plus the costs that occur to society. The social costs in the current

Table 5.4 *Botswana case examples*

Case example	Project	Baseline case	Indirect impacts included
Road pavement	Pavement of 531km sandy roads that link Botswana to Zambia, Zimbabwe and Namibia	Sandy roads with similar traffic level, but with 50% higher fuel consumption for the same traffic	Reduced SO_2 and NO_x emissions Increased employment in the construction phase
Efficient lighting	Introduction of compact fluorescent lamps, 11 watt	Incandescent lamps, 60 watt Coal-fired electricity	Reduced SO_2 and NO_x emissions
Industrial boilers	Improved coal-fired boiler with 85% efficiency by installing economizer on traditional boilers	Existing coal-fired boiler with 79% efficiency	Reduced SO_2 and NO_x emissions
Central PV plant	2 MW capacity of PV is established additional to the capacity that is assumed implemented in the baseline case	Coal-fired power production	Reduced SO_2 and NO_x emissions
Power factor correction	Installation of power factor correction	Coal-fired power plants without power factor correction Coal-fired electricity	Reduced SO_2 and NO_x emissions

project include net financial costs plus national benefits in the form of local air pollution reduction, and the value of time savings and employment generation.

The comparative assessment of the projects conducted with the CBA, the CEA and the MCA approach are based on the same quantitative data for the focal policy impacts as reported in Table 5.5.

As it can be seen from the table, the five projects considered are very different in scale: the road pavement project represents a large investment measured in gross as well as net financial costs, with a gross financial cost of about net present value (NPV) US$33 million, while the other projects have a gross financial cost that is not greater than around NPV US$3300. The NPV of the net financial costs are negative for all projects except for the central PV plant. This reflects the fact that the projects are expected to generate energy savings (for example, in the form of fuel savings) that have a greater value than the gross financial cost of implementing the projects. The road pavement project has been assessed to imply a

Table 5.5 *Project data for Botswana case examples[1]*

Case example	Gross financial cost NPV[2] US$1000	Net financial cost NPV[2] US$1000	Employment[3] (jobs created in man-years)	SO$_2$ emissions[4] tonnes	NO$_x$ emissions[4] tonnes	GHG reduction[4] tonnes
Road pavement	33,193	−257,489	1858	12,418	50,074	5,980,000
Efficient lighting	0.07	−1.0	–	34	11	11,600
Industrial boilers	3.0	−3.0	–	173	25	16,600
Central PV plant	3.3	2.5	–	959	368	87,496
Power factor correction	0.2	−0.06	–	157	60	15,038

Notes:
1 Measured as a change to a baseline case as specified in Table 5.4
2 Net present value over the lifetime of the project calculated with a 10 per cent discount rate
3 Accumulated over the lifetime of the project
4 Accumulated over the lifetime of the project

significant side-benefit in the form of employment generation, while the other projects, due to their limited scale, have been assessed not to have significant impacts on employment. All the projects will reduce local air pollution in the form of SO$_2$ and NO$_x$ emissions in addition to their capacity to reduce GHG emissions.

The project data that are reported in Table 5.3 have been used as input to six different technical assessments namely:

- A CBA with the decision criteria to maximize the total net economic impact of implementing the individual projects.
- A CBA analysis with the decision criteria to maximize the economic benefits of implementing the projects relative to the gross financial project costs.
- A CEA with the decision criteria to minimize the net financial project costs per unit of GHG emission reduction.
- A CEA with the decision criteria to minimize the social costs per unit of GHG emission reduction.
- An MCA analysis with the decision criteria to maximize the project score on the given policy indicators when equal weights are given to all policy impact areas.
- An MCA analysis with the decision criteria to maximize the project score on the given policy indicators. Particular high weights are given in this case to reduced SO$_2$ and NO$_x$ emissions while the other policy objectives are given lower weights.

Results of the CBA

Several different decision rules can be suggested for a CBA including 'rules' that prefer projects that maximize the total net benefits of the projects or rules that maximize the benefit/cost ratio for the projects.

The first decision rule considered here, where the objective is to maximize total net economic benefits, can be thought of as a relevant policy perspective for a developing country, given that the country expects to get financial compensation for the project implementation costs from international donors. The second decision rule that focuses on the benefit/cost ratio is a traditional economic investment perspective that often is used in the evaluation of public sector projects. The objective here is to maximize the payback on a public investment relative to other alternative investments.

The CBA by definition measures all policy impacts in welfare units, which in the current application are represented by monetary units. The values of reduced SO_2 and NO_x emissions are based on assumed avoided damage values,[9] and GHG emission reductions have been assigned a value of US$30 per tonnes of CO_2 equivalent emission units, which represent a simplified assumption about constant climate change damage per unit of avoided emission over time (see IPCC for studies on climate change damage values (IPCC, 2001b)).[10]

The results of the CBA in the form of total net economic benefit of the Botswana case projects are reported in Table 5.6.

The CBA rule as applied to the analysis reported in Table 5.6 implies the following internal project ranking:

1   Road pavement.
2   Central PV.
3   Power factor correction.
4   Efficient lighting.
5   Industrial boilers.

The project that generates the largest total benefit is the road pavement case, which is assessed to have a total net benefit of over US$500 million as reported in Table 5.6. Other projects that are assessed to generate large benefits are the central PV plant project and the power factor correction project, which are assessed to have net benefits of US$5.6 million and US$1.2 million, respectively. The result of the CBA analysis in the current specification particularly reflects the scale of the projects, where the two projects that represent the largest capital investment – namely, the road pavement and the central PV plant project – are the preferred ones.

Table 5.7 reports the results of an alternative CBA, where the overall policy objective is changed to be an evaluation of the net project benefits per unit of investment or 'benefit-cost' ratio.

The CBA as reported in Table 5.7 suggests that the ranking of projects according to maximum economic benefit per unit of investment is the following:

1   Efficient lighting.
2   Power factor correction.
3   Central PV plant.
4   Industrial boilers.
5   Road pavement.

Table 5.6 CBA-based analysis of Botswana case studies where the objective is to maximize the net economic benefits of the projects

Case example	Gross financial cost NPV US$1000 (1)	Net financial cost NPV US$1000 (2)	Value of employment NPV US$1000 (3)	Value of SO_2 emission reduction[a] NPV US$1000 (4)	Value of NO_x emission reduction[b] NPV US$1000 (5)	Value of GHG reduction[c] NPV US$1000 (6)	Net benefits NPV US$1000 (7)
Road pavement	N/A	−257,489	3169	82,734	15,637	229,101	588,130
Efficient lighting	N/A	−1.0	–	118.8	45.8	756.0	921.6
Industrial boilers	N/A	−3.0	–	205.0	18.0	393.0	619.0
Central PV plant	N/A	2.5	–	1640.4	638	3402	5681.9
Power factor correction	N/A	−0.06	–	372.9	144.5	663	1,179.5

N/A: Not applicable

Notes:
a Assuming a value of US$4.85 per tonne SO_2
b Assuming a value of US$4.90 per tonne NO_x
c Assuming a value of US$90 per tonne CO_2 equivalent
Column (7) is the sum of columns (2) through (6)

Table 5.7 *CBA analysis of Botswana case studies where the objective is to maximize the return on investments. Investment is measured as gross financial cost, and return is measured as the value of energy cost savings (fuel and O&M costs), employment generation, SO₂ and NOₓ emission reductions, and GHG emission reduction per unit of investment*

Case example	Gross financial cost	Energy cost savings per unit investment	Employment per unit investment	SO₂ emission reduction[a] per unit investment	NOₓ emission reduction[b] per unit investment	Value of GHG reduction[c] per unit investment	Total benefit per unit investment
	1000US$NPV (1)	US$NPV (2)	US$NPV (3)	US$NPV (4)	US$NPV (5)	US$NPV (6)	US$NPV (7)
Road pavement	33,193	8.8	0.1	2.5	0.5	6.9	18.7
Efficient lighting	0.07	15.3	–	1697.1	654.3	10,800	13,166.7
Industrial boilers	3.0	2.0	–	68.3	6.0	131.1	207.4
Central PV plant	0.2	1.3	–	1864.5	722.5	33515	5903.3
Power factor correction	N/A	–0.06	–	372.9	144.5	663	1179.5

Notes:
a Assuming a value of US$4.85 per tonne SO₂
b Assuming a value of US$4.90 per tonne NOₓ
c Assuming a value of US$90 per tonne CO₂ equivalent
Column (7) is the sum of columns (2) through (6)

This ranking of the projects reflects most crucially differences in the scale of the gross financial capital cost of the projects, where the highest priority is simply given to projects that have the smallest gross financial costs. In this way the most preferred projects become the lighting project and the power factor correction project.

Results of the Cost-effectiveness Analysis

The Botswana case projects have been assessed with two CEA, where the first alternative evaluates the net financial cost per tonne of GHG emission reduction and the second evaluates the social cost per tonne of GHG emission reduction.
     The results of the two CEA are reported in Tables 5.8 and 5.9.

Table 5.8 *CEA of the case projects based on net financial costs*

Case example	Net financial cost[a]	Value of employment	Value of SO_2 emission reduction	Value of NO_x emission reduction	GHG reduction transformed to	Net financial cost per GHG reduction
	US$1000 NPV (1)	US$1000 NPV (2)	US$1000 NPV (3)	US$1000 NPV (4)	US$ NPV (5)	US$ NPV (6)
Road pavement	−257,489	N/A	N/A	N/A	2545.6	−101.2
Efficient lighting	−1.0	N/A	N/A	N/A	8.4	−113.7
Industrial boilers	−3.0	N/A	N/A	N/A	4.4	−5.9
Central PV plant	2.5	N/A	N/A	N/A	37.8	67.1
Power factor correction	−0.06	N/A	N/A	N/A	7.4	−7.9

N/A: Not applicable
Notes:
a Including the value of energy savings
b The annual GHG emission reductions are transformed into an NPV, assuming that the avoided damages with GHG emissions are constant
     Column (6) is column (1) divided by column (5)

The ranking of the projects in the CEA that is based on net financial costs are the following:

1   Efficient lighting.
2   Paved roads.
3   Power factor correction.
4   Industrial boilers.
5   Central PV plant.

The cost-effectiveness of the projects here in particular reflects the relative magnitude of the energy-saving benefits generated by the projects compared with the gross financial project costs, and the efficient lighting project is attractive because it generates large electricity savings and has small gross financial costs. The road pavement project is also among the most attractive projects in this assessment because it generates a very high value of fuel savings. The results of the CEA based on the assessment of social costs per unit of GHG emission reduction are reported in Table 5.9.

Table 5.9 *CEA of the case projects based on social costs*

Case example	Financial cost (including value of energy savings)	Value of employment	Value of SO_2 emission reduction[a]	Value of NO_x emission reduction[b]	GHG emission reduction[c] transformed to	Social cost per tonne GHG emission reduction
	US$1000 NPV (1)	US$1000 NPV (2)	US$1000 NPV (3)	US$1000 NPV (4)	US$ NPV (5)	US$ NPV (6)
Road pavement	−257,489	3169	15,637	82,734	2545.6	−140.3
Efficient lighting	−1.0	N/A	118.8	45.8	8.4	−133.3
Industrial boilers	−3.0	N/A	5.0	18.0	4.4	−57.6
Central PV plant	2.5	N/A	1640.4	638	37.8	5.8
Power factor correction	−0.06	N/A	372.9	144.5	7.4	−78.1

Notes:
a Assuming a value of US$4.85 per tonne SO_2
b Assuming a value of US$4.90 per tonne NO_x
c The annual GHG emission reductions are transformed to an NPV, assuming that the avoided damages with GHG emissions are constant
Column 6 equals (column (1) − (2) − (3) − (4)) divided by column (5)

The decision rule that has been used in this version of the CEA is to prefer projects that have a low social cost per unit of GHG emission reduction. The ranking order of the projects according to this rule is:

1 Road pavement.
2 Efficient lighting.
3 Power factor correction.
4 Industrial boilers.
5 Central PV plant.

The project, which is based on social costs, is almost the same as the one based on net financial costs, except for the road pavement and the efficient lighting projects that switch in priority order in the two applications.

As we noted earlier, compared with the CBA, the CEA does not give priority to the 'scale' of the activity considered, which is the main reason for the different project rankings from those based on the CBA analysis. It also implies that, from the perspective of a country with no targets, the CBA may be more relevant.

Results of the MCA Assessment

The case projects for Botswana have been assessed finally with two alternative MCA approaches based on different weighting systems for valuation of the policy impacts. The results of these are reported in Tables 5.9 and 5.10. For each impact of relevance the table reports the value of $V(x)$ – ie, the normalized value of the impact, set on a scale from zero to one. The value of $W(x_i)$ is given by $V(x_i)$ multiplied by the weight. The value in the last column is the sum of the $W(x_i)$.

The use of a rule that gives equal weight to all policy objectives as in the MCA analysis that is reported in Table 5.10 means that the relative performance of the projects on each of the four policy objectives are considered as equally important. This implies that the value of a relatively high performance on one indicator, such as financial cost, is the same as the value of a relatively high performance on another area, such as SO_2 emission reductions. In this way, it is important to recognize that the project score on each indicator, $W(x_i)$, is established solely on the basis of the relative performance of the projects considered, which is different from the welfare-based weighting system that is used in the CBA analysis and the CEA analysis. These approaches assign 'absolute values' to the individual policy impacts and these can be based on market prices or estimated opportunity costs.

The MCA decision rule as presented here, then, is to maximize the sum of the indicator score on the objective function $W(x)$ per unit of GHG reduction, and the project ranking order accordingly becomes:

1   Road pavement.
2   Industrial boilers.
3   Power factor correction.
4   Efficient lighting.
5   Solar PV plant.

The MCA project ranking order is very similar to the ranking generated by the CEA based on social costs. The ranking order of the industrial boilers and the efficient lighting, however, switches between the two approaches. The similarity between the ranking in the MCA and the CEA reflects that almost equal weights are given to SO_2 and NO_x emission reductions in both of the approaches. This is the case because the opportunity cost values of the two pollutants are assumed to be almost similar in the CEA like assumed in the MCA. The industrial boilers move up in the MCA based ranking compared with the CEA, because the significant higher financial cost of the boiler option per unit of GHG reduction compared with the efficient lighting option, becomes less important in the MCA than in the CEA.

Table 5.11 shows an alternative MCA where higher weights are given to SO_2 and NO_x emission reductions, and lower weights are given to costs and employment generation.

Table 5.10 *MCA-based analysis of the case projects where all policy impacts are given equal weights*

Case example	Financial cost	Value of employment US$1000 NPV	SO_2 emission reduction[1] tonnes	NO_x emission reduction[2] tonnes	$W(x)^3$ Per unit of GHG reduction
Road pavement					
X_i/GHG	−101.10	1.24	0.002	0.01	0.65
$V(x_i)$	0.93	1.00	0.200	0.50	
Weight	0.25	0.25	0.250	0.25	
$W(x_i)$	0.22	0.25	0.050	0.13	
Efficient lighting					
X_i/GHG	−113.70	0.00	0.003	0.0009	0.34
$V(x_i)$	1.00	0.00	0.300	0.0500	
Weight	0.25	0.25	0.250	0.2500	
$W(x_i)$	0.25	0.00	0.080	0.0100	
Industrial boilers					
X_i/GHG	−5.90	0.00	0.01	0.02	0.51
$V(x_i)$	0.40	0.00	1.00	1.00	
Weight	0.25	0.25	0.25	0.25	
$W(x_i)$	0.01	0.00	0.25	0.25	
Central PV plant					
X_i/GHG	67.10	0.00	0.01	0.004	0.30
$V(x_i)$	0.00	0.00	1.00	0.200	
Weight	0.25	0.25	0.25	0.250	
$W(x_i)$	0.00	0.00	0.25	0.050	
Power factor correction					
X_i/GHG	−58.40	0.00	0.01	0.004	0.49
$V(x_i)$	0.75	0.00	1.00	0.200	
Weight	0.25	0.25	0.25	0.250	
$W(x_i)$	0.19	0.00	0.25	0.050	

Notes:
1 Total accumulated SO_2 emission reductions per unit of GHG emission reduction
2 Total accumulated NO_x emission reduction per unit of GHG emission reduction
3 The preference function is specific according to the definitions included in the section on the formalization of the decision problem

Table 5.11 *MCA-based analysis of the Botswana case examples based on 10 per cent weight to employment, 20 per cent weight to costs and 35 per cent weights to SO$_2$ and NO$_x$ emission reductions respectively*

Case example	Financial cost	Value of employment US$1000 NPV	SO$_2$ emission reduction[1] tonnes	NO$_x$ emission reduction[2] tonnes	$W(x)$[3] Per unit of GHG reduction
Road pavement					
X_i/GHG	–101.10	1.24	0.002	0.01	0.54
$V(x_i)$	0.93	1.00	0.200	0.50	
Weight	0.20	0.10	0.350	0.35	
$W(x_i)$	0.19	0.10	0.070	0.18	
Efficient lighting					
X_i/GHG	–113.70	0.00	0.003	0.0009	0.33
$V(x_i)$	1.00	0.00	0.300	0.0500	
Weight	0.20	0.10	0.350	0.3500	
$W(x_i)$	0.20	0.00	0.110	0.0200	
Industrial boilers					
X_i/GHG	–5.90	0.00	0.01	0.02	0.78
$V(x_i)$	0.40	0.00	1.00	1.00	
Weight	0.20	0.10	0.35	0.35	
$W(x_i)$	0.08	0.00	0.35	0.35	
Central PV plant					
X_i/GHG	67.10	0.00	0.01	0.004	0.42
$V(x_i)$	0.00	0.00	1.00	0.200	
Weight	0.20	0.10	0.35	0.350	
$W(x_i)$	0.00	0.00	0.35	0.070	
Power factor correction					
X_i/GHG	–58.40	0.00	0.01	0.004	0.57
$V(x_i)$	0.75	0.00	1.00	0.200	
Weight	0.20	0.10	0.35	0.350	
$W(x_i)$	0.15	0.00	0.35	0.070	

Notes:
1 Total accumulated SO$_2$ emission reductions per unit of GHG emission reduction
2 Total accumulated NO$_x$ emission reduction per unit of GHG emission reduction
3 The preference function is specific according to the definitions included in the section on the formalization of the decision problem

The weights used in the MCA analysis reported in Table 5.11 of the Botswana case examples, namely a 10 per cent weight to employment, a 20 per cent weight to costs and a 35 per cent weights to SO_2 and NO_x emission reductions respectively, generate another project ranking than the MCA analysis that was based on equal weights to all policy objectives as reported in Table 5.10. The project ranking with the new weights becomes:

1 Industrial boilers.
2 Power factor correction.
3 Road pavement.
4 Solar PV plant.
5 Efficient lighting.

The highest priority is now given to the industrial boilers, despite a relatively high cost per unit of GHG emission reduction. This is the result of significant reductions of SO_2 and NO_x emissions. For the same reason, the power factor correction project for the same reason moves up in priority. The road pavement project is only a third priority because the high energy cost savings of this project are given a relatively low weight in this assessment.

It should be noticed that the MCA in principle could be used to generate results that are similar to the results of the CEA that are reported in Tables 5.8 and Table 5.9. This can be done if the weighting rules of the MCA establish the same relative values of the policy objectives as the ones reflected in the prices or opportunity costs that are used in the CEA.

Discussion of the Differences in Results Generated by Different Approaches

The assessment of the five case projects for Botswana with different applications of the CBA, the CEA and the MCA approaches has resulted, as reported in the previous section, in different rankings of the projects. An overview of the different project rankings is shown Table 5.12. As can be seen from the table, there is a considerable variation in the project rankings across the different technical assessments and none of the five case projects has an identical ranking for all the assessments. This variation is not only seen in-between the assessment done with the different technical approaches, but also in different applications of each of the individual approaches.

The relative ranking of most of the projects varies considerably across the different assessments. Projects like the road pavement project, efficient lighting, and industrial boilers are assessed to be first priority projects in some of the assessments, but lowest priority in other assessments, and also the ranking of the central PV project varies considerably. Only the power factor correction project has a tendency to have relative similar priority in all the assessments.

The analysis so far has been based on those impacts that are amenable to quantification. As we noted earlier, this is important but not sufficient for a final decision on the priorities to be determined. The qualitative impacts also need to be assessed against the priorities for which they are relevant. The national climate change mitigation study for Botswana also included a qualitative evaluation of the projects that considered a number of the same policy objectives as the ones evaluated above (UNEP, 1999). The projects were qualitatively ranked in relation to the following policy objectives:

- accordance with government policy;
- implementability;
- impact on balance of trade;
- employment generation;
- social benefits;
- economic efficiency;
- benefits in other sectors;
- local environmental impacts.[11]

Table 5.12 *Ranking order of Botswana case projects based on alternative decision making approaches regarding the evaluation of financial and social costs, employment generation, SO$_2$-, NO$_x$- and GHG emission reductions*

	CBA Net benefits	CBA Benefit/cost ratio	CEA Gross financial costs	CEA Social costs	MCA Equal weights	MCA High weights to SO$_2$ and NO$_x$ emissions
Road pavement	1	5	2	1	1	3
Efficient lighting	4	1	4	2	4	5
Industrial boilers	5	3	1	4	2	1
Central PV plant	2	4	5	5	5	4
Power factor correction	3	2	3	3	3	2

On this basis the project that was given highest ranking in this assessment was efficient lighting, but the road pavement options and industrial boilers were also assessed to be attractive.

Some conclusions can be drawn from the case study where similar quantitative impacts of given projects have been evaluated using different technical methods of appraisal. It shows up that two major assumptions seem to drive the different results. The first of these is related to the specification of general policy objectives, where in particular the two alternative CBA assessments came out with very different results. The second set of assumptions relates to the valuation system or weighting rules that have been used to evaluate the different policy impacts. The implications of different valuation assumptions are seen most clearly in the two alternative MCAs. This is because the MCA in its basic structure does not specify any rules for the determination of weights and variations in these can imply a wide range of different project rankings. The potential variation of weights that are assigned to the policy impacts in a CBA and a CEA must be considered to be much more restrictive due to the welfare economic basis of these approaches.

The different applications of technical rules for decision-making imply a very different ranking of the case projects for Botswana, and it is therefore worth considering if this result suggests that the ranking of projects is purely a normative exercise that is legitimized by the technical assessments. Whether this conclusion holds depends on the structure of the decision-making process as such, including the initial selection of policy priorities, the application of these as part of objective functions for the technical assessment and the approach that is used to establish weights for the different policy objectives.

First, as previously stated, the determination of policy priorities and the establishment of an objective function for the technical assessment has a normative element, because there is no universal rule that suggests how to select development priorities that should be reflected in the specific context of the planning exercise. However, this normative element does not imply that the actual evaluation of project or policy performance in relation to a set of selected policy priorities has a normative character. One of the most important purposes of using a formalized technical assessment as part of the decision-making process is precisely to separate the first initial normative policy priority setting from the assessment of how different options meet these priorities, and in this way, as far as possible, to make an objective assessment of how the objectives can be met.

Second, the handling of trade-offs between different policy objectives requires the use of weights, and the establishment of these can imply that a normative element is introduced in the technical assessment. As outlined earlier, the welfare economic approaches like the CBA and CEA primarily establish weights on the basis of stated preferences on markets or on constructed markets using different methods for revealing preferences. In this way the idea of these approaches is to build on individual consumers' preferences that are stated through market transactions that are independent to the specific evaluation exercise. However, the use of constructed markets imposes a number of uncertainties and specific valuation problems, in particular for non-substitutable goods, as, for example, the loss of human life. The MCA uses a number of different approaches to the establishment of policy impact weights, including dialogues with policy-makers, or a broader audience and a large variety of weights can be included.

As concluded with regard to the selection of policy priorities, it is important, as far as possible, to separate the technical assessments and the establishment of policy impact weights in order to avoid the situation where the assessment is turned into a sophisticated procedure for justification of priorities based on predetermined policy choice.

LINKAGES BETWEEN POLICY-MAKING AND THE RESULTS OF TECHNICAL ASSESSMENTS

The evaluation of the critical assumptions implicit in technical assessments as exemplified in the GHG emission-reduction case projects for Botswana addresses a number of important issues of relevance to the implementation of individual projects in a broader public planning context. The scope of public planning, however, is much wider than reflected in these projects. Several options and combinations of options typically are considered in parallel, and implementation strategies will be developed in practice for more comprehensive national and sectoral programmes. The following section provides a brief overview of a number of approaches that can be used to broaden the perspective of the policy evaluation.

A public policy evaluation that includes, for example, GHG emission reduction as a policy objective most often will consider a range of options as given above, as well as a combination of these. A portfolio of options for the energy sector can include, for example, electricity end-use savings, efficiency improvements in the supply system, fuel substitution and various energy conversion technologies with low carbon intensity. The implementation of a combination of these options in some cases has different impacts from those that would have been expected if the options were assessed one by one. This can be the case when the options compete over scarce resources or when the performance of the options is interlinked. A number of examples of interlinkages are:

- GHG emission reductions achieved from energy savings depend on the fuel composition of the supply system.
- The costs of electricity savings options depend on the costs of the power system.

An accurate assessment of interlinkages between individual options like the ones just exemplified can only be done in integrated assessments, which are usually based on energy sector models. A number of GHG emission-reduction costing studies for the energy sector have been based on such models, including studies for industrialized countries as well as for developing countries (Kram and Hill, 1996; ALGAS, 1999; Halsnaes, Callaway and Meyer, 1998). These studies have emphasized not only the assessment of interlinkages between the energy supply and demand on the basis of detailed technology data, but the assessment of economic feedbacks from resource scarcities, for example through price effects, have been more limited in these models.

Several international programmes have developed approaches for the integration of several policy objectives in sectoral development programmes. An example of this is the World Bank Global Overlays Programme that is a framework for integrating GHG emission-reduction policies in general bank sector programmes. The programme includes guidelines for the energy sector and the forestry (World Bank, 1997). The Global Overlays are, by definition, policies that meet multiple policy objectives, including those related to general economic development, social issues and environmental policies, assuming that general World Bank sectoral policy priorities are fulfilled.

As we noted in Chapter 1, the United Nations Framework Convention on Climate Change (UN, 1992) has initiated the development of a number of international mechanisms for climate change finance where one of the more important ones is the Clean Development Mechanism (CDM) of the Kyoto Protocol. The aim of the CDM is to facilitate industrialized countries which, according to the Kyoto Protocol, have a commitment to reduce their GHG emissions, to offset this requirement by financing emission-reduction projects in developing countries (UNFCCC, 1997). According to Article 12 of the Protocol, it is a requirement that CDM projects assist sustainable development in the project host countries, but no actual specification of what sustainable development means in this context has yet been agreed by the parties to the Climate Convention (UN, 1992). One possible interpretation of the CDM sustainability clause is to suggest that the project impacts on broader development objectives should be evaluated with a planning approach like the one suggested in this chapter.

GENERAL CONCLUSIONS

This chapter has considered how sustainable development concerns can be addressed in an evaluation of policy options that both meet global objectives in the form of GHG emission reductions and local development objectives, including economic, environmental and social goals. The preferred approach is to use a public planning framework, where technical assessments of a range of policy objectives are undertaken, based on priorities that are established in an interactive process with policy-makers, or on the basis of stated preferences in national development programmes. Public planning in this context is understood as a multistage process that includes separate steps that identify policy goals, screen projects and policies, and transform goals into objectives that can be assessed in technical assessments, leading to an assessment of policy impacts which finally are used to generate information to the policy-making process.

The structure and critical assumptions included in a number of alternative technical approaches have been compared, and it was concluded that some of the most commonly used approaches like CBA, CEA and MCA exhibit a number of structural similarities. They are all based on an objective function that reflects a number of specifically chosen policy priorities and have in this way a normative character when applied to public planning problems. Another structural similarity in the approaches is that they all seek to integrate a number of different policy objectives into a more general decision-making framework, which in some cases will imply the use of weights to value individual objectives. It is suggested, as far as possible, to limit the number of trade-offs that are considered in the technical assessments through systematic analysis of the efficiency of different options and through eventually introducing 'safe minimum' standards for given areas.

A major difference between the welfare economic-based approaches such as the CBA and the CEA versus the MCA is the procedure that is used to establish weights for different policy objectives. The economic approaches primarily establish these weights on the basis of stated preferences of individuals in markets or in constructed markets, while the MCA tries to establish weights that reflect the relative performance of the policies and projects, which can be derived through a dialogue with policy-makers.

The differences and similarities between the different technical approaches were tested in an evaluation of five potential GHG emission-reduction options for Botswana. These approaches were used to establish information about project ranking with regard to the policy priorities of financial and social costs, employment generation and the reduction of SO_2, NO_x, and GHG emissions. It was concluded that two different versions of the CBA, the CEA and the MCA respectively lead to different project rankings. This result highlighted the importance of the specific weighting of individual policy impacts, as well as of the implicit political criteria on the results.

One conclusion drawn from this comparative study is the crucial role that a public planning exercise like the one considered here plays in separating the initial selection of general policy priorities from the subsequent evaluation of the performance of different options. This separation is essential in order to avoid the technical assessment becoming merely a sophisticated tool for the confirmation of pre-established policy priorities.

Another conclusion that can be drawn from the comparative assessment of the technical approaches is the importance of transparency in the reporting of the data inputs, the objectives and the objective weights. In the same way, final conclusions should be discussed critically in relation to the uncertainties that are inherit in such assessments, in particular with regard to the inclusion of impacts that are beyond the scope of conventionally marketed goods and services. Following that, it is recommended that sensitivity analysis of key parameters should be conducted and potential other assumptions that could lead to other conclusions should be discussed.

Finally, we should note that the weights and the choice between the technical approaches themselves may depend on whose perspective we take. For a developing country the weights will differ from those used by a donor or an industrialized country. Furthermore, if the developing country has no target, it may prefer to select its priorities based on CBA or MCA, rather than on the CEA approach.

NOTES

1 In this case the decision will be a reflection of sustainable development priorities that are relevant for the specific policy context.
2 A given objective function cannot integrate all relevant aspects of policy priorities, but can include, for example, a set of indicators that are selected to serve as representative pointers of the policy impacts considered.
3 The different approaches have been used in some cases to evaluate individual policy options, but have also been integrated in some studies in sectoral models.
4 It can itself be seen as an advantage that policy priorities that are used to represent development priorities are stated in a context that are delinked from the actual planning activity because the preferences in this way are expected to be 'unbiased'.
5 Informally, however, we have to take these into account. This is discussed further later in the chapter.
6 There are other technical requirements for an ordering such as reflexivity and continuity, but these two are the most important.

7   The impacts *x* can include the costs of the option (for example, inputs, such as labour and capital). If they are costs, the MWTP is that of the supplier of the goods who bears the costs. Under competitive conditions, these too will be equal to the prices of the inputs.

8   See Chapter 7 and UNEP (1999) for a detailed outline of all the assumptions applied to the assessment.

9   See Chapter 7 for more details on the 'benefit transfer' approach that have been used to establish values for avoided damages by reduced SO_2 and NO_x emissions.

10 Reduced GHG emissions have a value in the form of reduced climate change damages. Due to major uncertainties and the complex relationship between GHG emissions and climate change damages, it is not straightforward to estimate a reliable monetary value for avoided damages. The current CBA, therefore, should be understood only as an illustration of the impacts of giving the avoided GHG emissions a given value.

11 The reader will be aware that there is some overlap between this list and the impacts that have been quantified. Indeed, the line between what can be treated in the quantitative analysis and what cannot is not rigidly determined. Where, however, an impact appears in both analyses, care needs to be taken not to double count it.

REFERENCES

ALGAS (1999) *Asian Least-Cost Greenhouse Gas Abatement Strategy*, Asian Development Bank, Manila, the Philippines

Banuri, T, Weyant, J, Akumu, G, Najam, A, Pinguelli, R, Rayner, S, Sachs, W, Sharma, R, Yohe, G (2001) 'Scope of the report: Setting the stage: Climate change and sustainable development' in B Metz, O Davidson, R Swart, and J Pan (eds) *Climate Change 2001: Mitigation*, Contribution of Working Group III to the Third Assessment Report of the Intergovernmental Panel on Climate Change, Cambridge University Press, pp73–114

Bogetoft, P and Pruzan, P (1997) *Planning with Multiple Criteria, Investigation, Communication and Choice*, Copenhagen Business School Press, Denmark

Cantor, R and Yohe, G (1998) 'Economic activity and analysis' in S Rayner and E L Malone (eds) *Human choices and Climate Change*, Vol 3, The Tools for Policy Analysis, Battelle Memorial Institute, Pacific Northwest National Laboratory, Seattle

Daly, H E (1990) 'Toward some Operational Principles of Sustainable Development', *Ecological Economics*, Vol 2, pp1–6

Halsnaes, K, Callaway, J M and Meyer, H (1998) *Methodological Guidelines*, Economics of Greenhouse Gas Limitations, Main Report, Risø National Laboratory, UNEP Collaborating Centre on Energy and Environment, Denmark

IPCC (2001b) *Climate Change 2001: Impacts, Adaptation, and Vulnerability*, Contribution of Working Group II to the Third Assessment Report of the Intergovernmental Panel on Climate Change, Cambridge University Press, Cambridge

Jasanoff, S and Wynne, B (1998) 'Science and decisionmaking' in S Rayner and E L Malone (eds) *Human Choices and Climate Change*, Vol 1, The Societal Framework, Battelle Memorial Institute, Pacific Northwest National Laboratory, Seattle

Johansen, L (1979) *Lectures on macroeconomic planning, 1. General aspects*, Elseviers North-Holland, New York

Keeney, R L, Raiffa, H (1993) *Decisions with Multiple Objectives. Preferences and Value Tradeoffs*, Cambridge University Press, Cambridge

Kram, T and Hill, D (1996) 'A Multinational model for CO_2 Reduction: Defining Boundaries of Future CO_2 Emissions in Nine Countries', *Energy Policy*, Vol 24, No 1, pp39–51

Markandya, A, Halsnaes, K (co-lead authors) and Lanza, A, Matsuoka, Y, Maya, S, Pan, J, Shogren, J, Seroa de Motta, R, Zhang Tianzhu (lead authors) (2001) 'Costing Methodologies', Chapter 7 in *Climate Change 2001: Mitigation*, Contribution of Working Group III to the Third Assessment Report of the Intergovernmental Panel on Climate Change, Cambridge University Press, Cambridge, pp451–498

Markandya, A, Halsnaes, K and Milborrow, I (1998) 'Cost Analysis Principles' in K Halsnaes, J Sathaye and J Christensen (eds) *Mitigation and Adaptation Cost Assessment: Concepts, Methods and Appropriate Use*, Risø National Laboratory, UNEP Collaborating Centre on Energy and Environment, Denmark

Pearce, D W, Barbier, E W and Markandya, A (1990) *Sustainable Development*, Earthscan, London

Pezzey, John (1992) *Sustainable Development Concepts, An Economic Analysis*, World Bank Environmental Paper No 2, Washington, DC

Ray, A (1984) *Cost Benefit Analysis: Issues and Methodologies*, The Johns Hopkins University Press, Baltimore

Roseland, Mark (2000) 'Sustainable community development: integrating environmental, economic and social objectives', PROGRESS IN PLANNING 54 (2000), pp73–132

Sathaye, J, Norgaard, R and Makundi, W (1993) 'A Conceptual Framework for the Evaluation of Cost-Effectiveness of Projects to Reduce GHG Emissions and Sequester Carbon, LBL-33859', Lawrence Berkely Laboratory, US

Squire, L and van der Tak, H (1975) *Economic Analysis of Projects*, The Johns Hopkins University Press, Baltimore

Toth, F, Mwandosya, M (co-lead authors) and Carraro, C, Christensen, J, Edmonds, J, Flannery, B, Gay-Garcia, C, Lee, H, Meyer-Abich, K M, Nikita, E, Rahman, A, Richels, R, Ruqie, Y, Villavicencio, A, Wake, Y, Weyant, J (lead authors) (2001) 'Decision-Making Frameworks' in *Climate Change 2001: Mitigation*, Contribution of Working Group III to the Third Assessment Report of the Intergovernmental Panel on Climate Change, Cambridge University Press, pp601–688

UN (1992) *United Nations Framework Convention on Climate Change*, International Legal Materials, Vol 31, 1992, pp849–873

UNEP (1999) *Economics of Greenhouse Gas Limitations. Country Study Series Botswana*, UNEP Collaborating Centre on Energy and Environment, Risø National Laboratory, Denmark

UNFCCC (1997) *Kyoto Protocol to the United Nations Framework Convention on Climate Change*, (UNFCCC), FCCC/CP/1997/L.7/Add.1, Bonn

World Bank (1997) *Guidelines for Climate Change Global Overlays*, Global Environment Division, paper no 047

Chapter 6

Sustainability in Climate Mitigation: Integrating Equity into Project Analysis

Tim Taylor[1]

INTRODUCTION

Climate change is likely to have significant impacts on poor people around the world. The poor are less able to adapt to change and are often most at risk from environmental degradation as a result of this lack of adaptive capacity or because they are most exposed due to their location. Distributional concerns on climate change should not be limited, however, to the impacts, as climate mitigation and adaptation measures are likely to have significant direct and ancillary impacts on people in different income groups and across generations.

The impact of climate mitigation policy on sustainable development has been discussed to some extent in Chapter 2, which highlighted the need for the consideration of implementation barriers in the assessment of climate mitigation policies. One such potential barrier is given by the distribution of the burden of mitigation across different sectors of the economy and across income groups. Previous chapters have discussed the issues of the impact on social capital of the evaluation of climate change projects (Chapter 4) and analytical approaches for the evaluation of mitigation projects (Chapter 5). This chapter builds on the analytical framework developed in Chapter 5, highlighting techniques by which the crucial issue of the distribution of impacts may be considered in this context.

This chapter is structured as follows. First, the impact that distributional considerations have on sustainability is discussed. This section reviews both the inter- and intragenerational equity debate. The focus then turns to intragenerational equity, with an examination of the key measures of intragenerational equity. The costs, benefits and ancillary benefits of climate change mitigation policies are then examined, with particular reference to the distributional nature of such measures. A methodological framework for the inclusion of equity impacts in the analysis of mitigation policies or projects is then presented. The chapter concludes with an applied example of this framework, drawing on the Botswana case study presented in Chapter 7.

DISTRIBUTION AND SUSTAINABLE DEVELOPMENT

The issue of equity is central in moves towards sustainable development. Countries with a higher level of equity have been shown to have higher growth rates compared to those with relatively greater inequality.[2] While intergenerational equity has been the focus of much debate, with discussions focusing on the appropriate intertemporal discount rate, the issue of intragenerational equity has largely been put to one side in the climate change debate. However, it is increasingly recognized that matters of intragenerational equity may be of more importance to national decision-makers for whom current socioeconomic problems are of greater importance than future troubles. It is in this context that this chapter attempts to draw together the literature and present a framework for the evaluation of climate mitigation projects.

Intergenerational Equity Issues

The distribution of the impacts of mitigation projects across generations is clearly linked to sustainable development. The issue of intergenerational equity in climate change project assessment has been widely discussed in the literature, with the use of discount rates being the main focus of discussion. Some, notably Lind (1995), have questioned the use of cost-benefit analysis for the climate policy case. This section discusses the different discount rates that have been suggested in the literature.

Different discount rates and techniques have been suggested to highlight the issue of intergenerational equity and the impacts of climate change. The IPCC suggests two broad categories of approach to discounting: the prescriptive and descriptive approaches (1996, 2001). The prescriptive approach takes estimates of the absolute value of the elasticity of marginal utility and the growth rate of per capita consumption, assuming that the rate of pure time preference takes a value of zero.[3] This approach yields discount rates from the order of 0.5–16 per cent, depending on the country specific growth rate and the elasticity of marginal utility, which may be high in poor countries under assumptions of rapid increases in utility from movement out of deprivation. The descriptive approach examines the marginal rate of return on capital, which may be 5–10 per cent, and sets this equal to the discount rate owing to the potential for displaced investment if this not used.

As an alternative to the standard, neoclassical discounting which assumes a constant exponential rate, hyperbolic discounting has been suggested for potential application in project evaluation (Cropper and Laibson, 1999). Hyperbolic discounting involves the use of higher discount rates in evaluating near-term impacts than are used in long-term impacts. Cropper and Laibson suggest 'there is. . . strong empirical evidence that people discount the future hyperbolically'. However, the use of hyperbolic functions, and as a consequence lower discount rates, is not suggested by Cropper and Laibson only for environment-related projects, though this may be the case under certain conditions.[4]

The debate on discounting and the implications for climate policy is a longstanding one and will not be simply resolved. Cline (1999) argues that it is necessary to move beyond traditional discounting techniques, which are mostly applicable for short-term projects. Weitzman (1998) surveyed 1700 economists and found that the majority believed that differential discount rates were needed for environmental problems with a long-time horizon, such as climate change. The discount rate implied by Weitzman's analysis falls progressively from 4 per cent in the present to 0 per cent for the far distant future. Kopp and Portney (1999) argue for the use of 'mock referenda' where households would be asked to state their preferences on long-term policy issues, such as climate change, based on carefully designed questionnaires. Problems exist, however, in the application of this technique, following those identified for the contingent valuation method, particularly in the presentation of the policy choice. The issue of the correct application of cost-benefit analysis to the climate change policy decision is complex, and is beyond the scope of this chapter. The impacts of climate change policy are hard to estimate, given uncertainties – hence the project evaluation conducted in Chapter 7 and later in this chapter focuses on project costs and ancillary benefits related to the projects. Thus it is a near-term policy choice, assuming that policy is needed to combat climate change. The main equity concern in the near term is that of intragenerational equity, which may have increasing importance in moves towards a sustainable climate policy framework, and this will be the focus of the rest of this chapter.

Intragenerational Equity

Intragenerational equity – the distribution of income and other factors that make up well-being (including environmental quality) within a generation – is important for the sustainability of the development process and hence of the mitigation strategy. The impact of equity on two of the determinants of a sustainable policy have to be considered: the importance of equity to development itself and the importance of equity to the successful implementation of policy.

Equity may be an important development objective for governments. In particular, the reduction of poverty through income redistribution is generally a key part of government policy. Hence mitigation projects with a progressive impact are likely to be favoured over those with a regressive impact. It may also be the case that some mitigation projects are undertaken primarily as development projects. These projects would include those which improve the potential productivity of the workforce, such as the energy-efficient lighting scheme suggested in Chapter 7, which may have gains for education and thus meet important development objectives that may be of higher priority than climate change mitigation to the national government.

The degree of inequality in a country may have a number of potential impacts, including increased social unrest and slower economic growth.[5] It may also have important implications for the level of environmental degradation.[6]

The distribution of impacts resulting from climate change mitigation may also have important implications for the political feasibility of the project. Where the impacts focus on a politically active group or on an important interest group in society, the implementation of such a policy may be stalled and the policy rendered unfeasible. Thus, even if a policy has a negative economic cost and a low financial cost, if there are negative impacts on a key interest group, this may affect the feasibility of implementation in the eyes of the policy-maker. This problem was highlighted by Dixit (1996) who illustrates that, even in the case where compensation of the affected group is possible, credibility issues may render the policy impossible to implement.

An example can be given based on the Zimbabwean and Botswana case studies presented in Chapter 7. As coal-miners in Zimbabwe and Botswana are politically active and are an important interest group in determining voting patterns, contributions to political parties and the like, this would be of crucial importance in determining the favoured mitigation policies in these countries. The acceptance of specific policies by the miners, however, is a complex issue, and cash compensation and/or redeployment for negative impacts is not a guarantee of acceptance. How they are affected by the mitigation option is of great importance. Dixit illustrates the importance of this in a transactions-cost approach to political economy as follows:

> *The workers know that their receipt of compensation in the future will depend on their future political power, not on the economic choices they make to stay in the declining industry or move to a new industry or location. In fact they may suspect that moving will disperse their forces and reduce their political strength. Therefore their incentives to relocate to more productive opportunities are blunted and they may use their political power to demand and get help in their current occupation.*

Clearly, therefore, the assessment of the distributional impacts on stakeholders and by stakeholders has important implications for the potential implementation of a mitigation project.

Previous work on the integration of distribution into project analysis includes Xu (1994) which focused on the issue of the distributional impacts of forestry projects, within the context of sustainable development. Xu established a systematic approach to this analysis, which provides some insight into the processes through which one has to go before presenting a distributional impact assessment. The overall framework, adapted to the issue of climate change mitigation, is presented in Figure 6.1.

As shown in the figure, there are several steps that must be taken to assess the distributional impact of policies. First, one must identify the factors that have distributional impacts. These may be physical, technical, social, economic or institutional. The impacts are valued in money terms where possible. Groups of individuals are identified so that the impacts may be assessed as to the differential impact on different groups. These groups may be based on income levels, social class or other classification systems. A summary measure of the impacts on different groups is calculated in order to identify the impact on inequality of the given measure and to allow easy comparison of different policies. Intragenerational equity, the distribution of income and other factors that make up well-being (including environmental quality) within a generation, is important for the sustainability of the development process and hence of the mitigation strategy. The impact of equity on two of the determinants of a sustainable policy have to be considered: the importance of equity to development itself and the importance of equity to the successful implementation of policy.

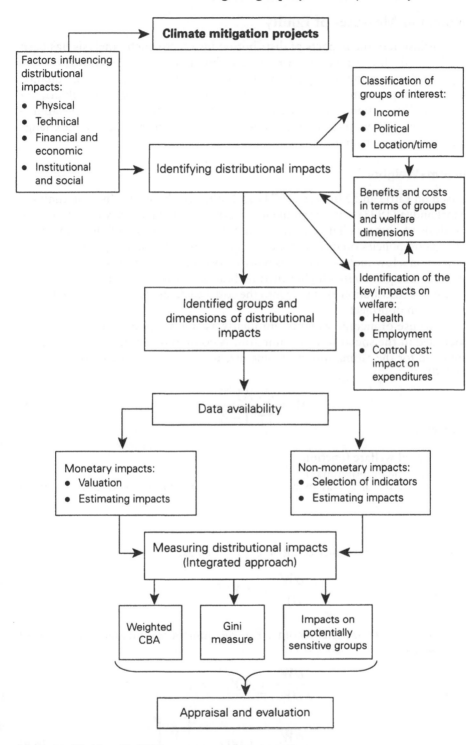

Figure 6.1 *Framework for distributional impact assessment*

Source: Modified from Xu (1994)

Common Measures of Equity

The inclusion of the impacts of distribution into cost-benefit analysis has been applied in the literature using three main techniques: income weighting, averaging and classification of impacts by individuals. Other measures of inequality, such as changes in the Gini coefficient, may also be used as part of multicriteria analysis. This section will present a review of the literature to date on the application of these techniques in the analysis of projects in general and climate change mitigation projects in particular.

Income Weights

Based on Little and Mirrlees (1974) and other work on benefit-cost analysis, Markandya (1998) considered the impact of including income weights into the analysis of climate change mitigation projects. Weighting of the impacts of mitigation policies on different income groups is based on the premise that an impact on a low-income person is worth more, in welfare terms, than the same impact on a person in a higher income bracket. This technique integrates the concept of distribution explicitly into the cost-benefit analysis framework for the evaluation of projects.

The use of income weights[7] implicitly assumes that income inequality is included in the social welfare function. A commonly applied social welfare function for the estimation of income distribution weights is that of Atkinson (1970):

$$W = \sum_{i=1}^{N} \frac{A Y_i^{1-\varepsilon}}{1-\varepsilon} \tag{6.1}$$

where:

W: social welfare function
Y_i: income of individual i
ε: elasticity of social marginal utility of income or inequality aversion parameter

A: a constant.

The social marginal utility of income is defined as:

$$\frac{\partial W}{\partial Y_i} = A Y_i^{-\varepsilon} \tag{6.2}$$

Taking per capita national income, $\bar{Y}$ as the numeraire, and giving it a value of one, we have:

$$\frac{\partial W}{\partial Y_i} = A \bar{Y}_i^{-\varepsilon} = 1 \tag{6.3}$$

and

$$\frac{\partial W}{\partial Y_i} = SMU_i = \left[\frac{\bar{Y}}{Y_i} \right]^{\varepsilon} \tag{6.4}$$

Thus, the social marginal utility of one unit of income received by individual i (SMU_i) is taken to be the proportion of average per capita income to the individual's income raised to the power of an income inequality aversion parameter, ε.

The distributional weight suggested above takes average income as the numerator. Little and Mirrlees suggested the use of a 'critical consumption level' as the numerator. The use of consumption, rather than income, highlights the issue that is common in the distributional literature of the distinction between annual and lifetime income.

The use of distributional weights in project appraisal has been the subject of considerable debate. Arguments against the use of distributional weights include the so-called 'arbitrary' nature of weighting, although the weights can be derived from value judgements and estimates of the inequality aversion parameter have been obtained from observation of consumption patterns and government taxation policies (see below).

Some practical problems exist in the determination of the distribution of the costs and benefits of mitigation policies. This requires household expenditure surveys to estimate the impacts the of increased costs of energy to different groups and population census statistics to evaluate the distribution of the ancillary benefits derived from reduced air pollutants. The exercise becomes more complex if the direct recipient of the gain is not the ultimate recipient of the gain. For example, improvements in environmental quality may increase property price, leading to some of the gain being allocated to landlords. Thus, the distribution of ancillary benefits from lowered air pollution may not be as clear-cut as first thought.[8]

As noted above, the use of distributional weighting has been criticized as allowing for the arbitrary application of weights to impacts and having an adverse effect on project selection as a result. It is suggested by those who argue against weights that compensation can be built into the project design and that this can reduce the distributional problem. If compensation is built in, however, this does not remove the need for an assessment of the distributional impact. Such an assessment is needed to confirm the effectiveness of the compensation mechanism.

It can be argued that a weighted cost-benefit analysis may enable a better evaluation of projects than is possible under a multicriteria assessment, highlighting key areas of distributional concern, where projects need to be modified. The use of weights which reflect previous policy decisions can assist in this and it is the determination of the inequality aversion parameter which determines the weight applied, to which we now turn.

Determination of the Inequality Aversion Parameter

The determination of the inequality aversion parameter ε is critical for the application of distributional weights to the impacts of projects and policies. For developing countries, the negative consequences of inequality may be greater, leading to a greater value of ε than in developed countries. This is borne out by a review of estimates for Atkinson's inequality aversion parameter. Table 6.1 presents some estimates for the value of this parameter that have emerged from the literature. For India, an inequality aversion parameter of 1.75–2.0 was found by Murty et al (1992). Estimates for the inequality aversion parameter for the UK were reviewed in Stymne and Jackson (2000). The value of ε ranges from 0.5 to 10.0, with more precise estimates in the range 0.5–0.8, suggesting that inequality aversion in the UK is lower than that in India. On the basis of this, different inequality parameters need to be applied in different countries. For developed countries, sensitivity within the range of 0.5–1.0 may be appropriate, whereas for developing countries the aversion parameter is likely to fall in the range 1.0–2.0.

Table 6.1 *Estimates of inequality aversion*

Country	Estimate	Study
India	1.75–2.0	Murty et al (1992)
Kenya	1	Scott et al (1976) (cited in Brent (1990)
Norway	0.8	Christiansen and Jansen (1978) (cited in Brent (1990))
UK	0.5–0.75	Shwartz and Winship (1979)
UK	0–10 (most likely 2.0)	Stern (1977)
UK	0.8	Blundell, Brewning and Meghit (1994)

Source: Based on Brent (1990), Murty and Markandya (2000) and Stymne and Jackson (2000)

Brent (1990) presents two types of aversion parameter that have been used in the establishment of inequality weights: a priori parameters and parameters that have been imputed from past government policy on taxes and consumption. The use of a priori parameters involves the explicit specification of weights before analysis is conducted of the various policy options. These weights may be based on the expert's philosophical opinion or introspection. The parameter ε may be determined as that which the individual would choose under uncertainty (when they did not know their position in society). It has been suggested, however, that such a method would result in setting a value of ε equal to infinity (following the maximin principle of Rawls (1971)). In general, however, the arbitrary fixing of distributional weights such as may result from the use of a priori estimates of ε has been criticized for its potential to overemphasized distributional issues.

Imputing weights from observation of past policy decisions, such as observation of the tax system (Stern, 1977; Christiansen and Jansen, 1978) or consumption patterns provides a less arbitrary source of estimates for ε. Past policy reflects a government's interest in distributional issues and thus can be used as a source for determining future policy. This assumes that past policy took full account of distributional concerns. Scott et al (1976) imputed a value of 1 for ε in Kenya based on the distributional incidence of income taxation, school fees and the barely progressive nature of the tax system in rural areas.[9]

Gini Coefficient and Measures of Inequality

The most commonly used measure of inequality is the Gini coefficient that measures the relative distribution of income across different deciles of the population. This measure is attributed to Gini (1912). The Gini coefficient (G) measures the area between the Lorenz curve and the line of equity (shown in Figure 6.2). This can defined as follows:

$$G = 1/2\mu^* \Sigma |x_i - y_i| \qquad (6.5)$$

Where μ represents the average income level, x_i represents the income held by group i, and y_i the income that would have to be held for equality. Thus it is a measure of the absolute differences between the income levels of different percentiles in the region and the levels of income they would expect under equality. The Gini coefficient lies between 0 and 1, with 0 indicating equality and the values approaching 1 indicating that there is a high level of inequality.

The use of the Gini measure in project evaluation is complicated by the fact that many projects will have only marginal or minute impacts on this measure. The Gini measure has also been shown to have some important methodological flaws (see Atkinson (1970), Sen (1997)), but remains a common measure of inequality.

Poverty Measures

Climate change mitigation projects are likely to have diverse impacts on different income groups and measures of the impacts on the degree of inequality in the society are presented above. Inequality may be included explicitly in a cost-benefit analysis framework, such as income weighting, or as part of multicriteria analysis, as in the case of a change in the Gini measure. However, a change in an inequality measure does not necessarily imply a change in the level of poverty, which is a major issue in policy in developing countries. Feldstein (1998) argues

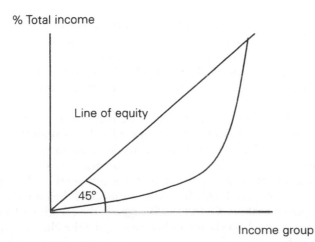

Figure 6.2 *Lorenz curve*

that the focus on inequality reduces the potential for the consideration of impacts on poverty, which may be a far more important policy objective.

The development of a measure of the impact on poverty is necessary as the relative distribution of impacts of climate change mitigation projects, as highlighted in the section above, may fail to take fully into consideration the implications on poverty of such actions. This framework may be of particular use to the international agency in determining projects that most fit with poverty alleviation objectives, and enable some cross-country comparison of the poverty alleviation potential of given climate change mitigation projects.

The Asian Development Bank has developed a measure of the relative impact of the benefits of projects on poor groups in society, the poverty impact ratio (ADB, 2001). This ratio takes a measure of the relative proportion of benefits accruing to the poor, which can then be compared to the proportion of the population in that group to assess the impact on poverty of a given project.

Identification of Impacts on Key Individuals or Groups

In an extension to either of the above techniques for the integration of equity issues into the analysis of projects, or as a free-standing exercise as part of a multicriteria analysis, the impacts on certain key groups in society can be identified. Although not strictly an equity issue, the impact on key groups in society may be important in the implementation of projects,[10] where groups are important in the political process or where impacts on such groups are of importance due to other policy commitments. This was identified as an important issue affecting the implementation of mitigation policy in Chapter 2.

Key groups may be determined according to their contribution to GDP, their level of participation in the political process, ethnicity (eg where negative impacts on certain ethnic groups may lead to tension between that group and the state, such as is the case in Malaysia (Bourguignon, de Melo and Morrisson, 1991)).

One possible classification of stakeholders is presented in ODA (1995). This classification divides stakeholders into those that are 'primary', 'secondary' and 'key':

- Primary: those ultimately affected by the option, positively or negatively.
- Secondary: those involved in the delivering of the option, including those involved in the decision-making and those excluded.
- Key: those who may be indirectly affected by the option, but who may exercise a large degree of influence that can affect the intervention.

The importance of stakeholder analysis to the implementation of policy is highlighted in Taylor, Markandya and Hunt (2001). Stakeholders may be divided between those whose views may be considered important to the policy-maker and those who have a large degree of influence over the policy-setting agenda. Such a classification aids the analysis of the potential implementation of climate mitigation policy, as it highlights the importance of the impacts on certain stakeholders to policy-makers. ODA (1995) suggests the creation of a matrix to facilitate the analysis, such as shown in Figure 6.3, for the climate policy case. This also follows the ODA classification of stakeholders shown above.

A detailed analysis of the impacts of climate mitigation policy on the groups identified in such a framework should be conducted in order to facilitate the decision-making process. On one hand, the identification of impacts and the level of importance attributed to the different actors is needed in order that the measures may be designed to offset some of the costs. On the other hand, such analysis should precede policy actions as they are important in terms of the successful implementation of policy, particularly where stakeholders are important in the political process.

Costs, Benefits and Ancillary Benefits of Mitigation Policies

Having evaluated the techniques used for the inclusion of values and measures of intragenerational equity into the analysis of projects, we now turn to the matter of the estimation of the control costs direct benefits and ancillary benefits of mitigation policies. The determination of the distribution of these benefits and the appropriate weighting of such impacts is necessary for the ranking of mitigation policies.

	Low influence	**High influence**
High importance	1, 2 3	4, 5, 6, 9
Low importance		7, 8, 9

Notes:

Primary stakeholders:
1 = poor households
2 = household industries
3 = richer households
4 = mining industry
5 = workers' unions

Secondary stakeholders:
6 = national government
7 = UNFCCC

Key stakeholders:
8 = environmental NGOs
9 = energy producers (influence dependent on national conditions)

Figure 6.3 *Stakeholder analysis matrix for climate policy case*

Distribution of Control Cost

The distribution of the control costs resulting from GHG mitigation projects depends on the method used to raise the revenues to fund the given project. In general, mitigation projects may fall into three main categories: self-funding; funded through government taxation; and funded from an external source. There may be some degree of overlap between these categories, and where that occurs the distribution should be assessed accordingly. This section will examine techniques that may be or have been applied to assess the implications of each on the distribution of control costs.

Self-funding Measures

Self-funding measures are those that have the potential for revenue collection to cover costs. Such cost-recovery projects may be funded through additional charges on services – for example, road tolls or charges on vehicle inspections. The distribution of the willingness to pay for services needs to be assessed and the cost implications on different income groups of applying a charge needs to be considered. It may be necessary to provide, through the design of the charge scheme, some degree of protection for those most vulnerable to rising costs.

Taxation

Mitigation strategies may require some central government funding. Different types of central government taxation have different implications for distribution and thus require separate estimates of the welfare losses to different income classes.

Carbon taxes, which are one potential means of mitigating against climate change, has been studied from a distributional perspective, along with other environmental taxes. Several papers to date have focused on the distributional implications of the costs of such taxation. Through the examination of consumption patterns, most have concluded that the cost of such taxation is regressive in nature, as environmentally harmful goods are consumed disproportionately by poorer sections in society. This analysis has been largely based, however, on data from developed countries, such as Australia (Cornwell and Creedy, 1997) and the US (Metcalf, 1998) and few studies have examined developing countries.

To date, most of the work on environmental taxation has focused on consumption patterns and the implications of raising taxes on goods with an environmental element on different sectors in society. The degree of sophistication in these estimates varies, with some examining just the impact of raising taxes and assuming that expenditures remain constant, and others examining more flexible consumption patterns using elasticities to estimate changes in consumption arising from the imposition of environmental taxes (Cornwell and Creedy, 1997). Other studies have considered the impact of raising carbon taxes, which would have an impact not only on the consumption side of the economy but also on prices of goods which use taxed commodities in their productive processes, drawing on input–output models (Cornwell and Creedy, 1997; Shah and Larsen, 1992).

Eskeland and Kong (1998) analysed the distributional impacts of what they term 'control strategies' for various sectors in Indonesia. This analysis did not include the potential benefits arising from the use of revenues from environmental taxation. Instead they assessed the distributional impacts of various 'control strategies' through an examination of the burdens on different income groups, using consumption shares of different income groups. The analysis also included some analysis of the benefits of environmental improvement, drawing on a willingness to pay and a sensitivity analysis of the impacts using different income elasticities for willingness to pay. Their findings are that private transport control distributes costs more progressively among households than a public transport strategy, and both are more progressive than controls on energy use. This is consistent with the expenditure shares, where poorer groups spend a larger proportion of household expenditure on energy compared to richer groups. For transport, more is spent by low-income families on public transport than on private transport; hence controls on public transport affect the poor more than those on private transport.

Cornwell and Creedy (1997) present a framework for the evaluation of the distribution of the costs of environmental taxation, drawing on the Australian Household Expenditure Survey. Using the assumption of additive preferences, they derive estimates for the elasticities of expenditure on different commodities for different income groups, building on the work of Frisch (1959). These elasticities were used in the estimation of the distributional effects of imposing a tax on household fuel and for analysing the impacts of the imposition of a carbon tax. In evaluating the impacts of imposing a carbon tax, the analysis was extended through the use of input–output techniques to include impacts on the prices of commodities which use carbon-containing substances in their productive processes. The main findings of this study were that a 30 per cent tax on fuel would have a slightly less negative effect on inequality than a 15 per cent tax on food. Carbon taxes were found to increase inequality slightly in the Australian case, although the distributional consequences were less than those estimated by Symons, Proops and Gay (1994) for the UK. An interesting finding was that transfer payments could be adjusted in order to mitigate the regressive impacts of both carbon and fuel taxes without decreasing overall revenue.

The distributional impact of raising taxation or prices of energy is an important issue. Similarly, in project analysis where the projects in question involve some state funding, the source of this funding needs to be investigated to assess the distributional impact of the project as a whole.

The same techniques as used in the studies for environmental taxation could be used to assess the impacts of new control costs on energy as part of a multi-criteria analysis of a project or as part of an extended cost-benefit analysis. Here, the impact on such measures of inequality as the Gini coefficient and other measures commonly used in the analysis of distribution could be provided to the policy-maker to better inform the decision-making process. However, impacts on inequality from the cost side can also be included by extending the conventional cost-benefit analysis of a given set of mitigation projects through the use of income weights, whereby a weight is assigned to impacts according to the nature of the distributional effect of the measure. This technique was used in a cost-benefit analysis of the Ganga Action Plan, which assessed the impacts of cleaning the River Ganges (Murty and Markandya, 2000).

Distribution of Direct Benefits of Mitigation

The estimation of the distributional impacts of climate change, and hence the avoided damage as a result of the implementation of mitigation strategies, has proved difficult. While it is generally accepted that developing countries and the poorer groups in society have less adaptive capacity and would be more heavily affected by climate change, the valuation of this difference is open to some debate. Changes in lifestyle, the changes in climate attributable to increases in GHGs and future technologies mean that the estimation and evaluation of the direct benefits of mitigation is complex (Schelling, 1992). Further, issues of discounting loom large, with some advocating that growth may potentially mean that the burden of climate change is substantially less than would be expected (Cline, 1993; Lind, 1995). Given the difficulty in measurement and the marginal nature of emissions relative to overall climate change, the analysis of the distributional implications of mitigation strategies contained in this chapter will not consider the direct benefits of mitigation. More important is the valuation of ancillary benefits and it is to this issue that we now turn.

Distribution of Ancillary Benefits

The ancillary benefits of climate change mitigation projects include:

*   Reductions in health damages as a by-product of reduced air pollutant emission;
*   Increases in employment, resulting in health improvements;
*   Increased educational prospects; and
*   Reduced noise impacts.

A fairly extensive literature on several of the main ancillary benefits exists, particularly with relation to air pollution. The extent to which poorer groups in society are exposed to higher levels of pollution have been examined in some studies for both developing and developed countries. A brief review of this literature follows, focusing on the air and employment impacts of policy measures.

This study will focus on the distribution of air pollution and employment effects, as these have been most quantified in the literature. These are also likely to be most significant in terms of priorities for policy-makers.

Distributional Impacts of Air Pollution

Markandya (1999) presents a review of some of the literature on the poverty–environment–development linkage, generally finding that the impact of poverty on the environment and vice versa is mixed. For urban areas, it may appear that the distribution of environmental degradation is overwhelmingly regressive, with poorer areas suffering from higher concentrations of pollutants. However, evidence from the US (Brooks and Sethi, 1997) suggests that this relationship might not be clear-cut, with poor and rich living side by side in highly polluted areas. Despite this, one can infer that the poor are generally more prone to experiencing damage as a result of environmental degradation, as they are less able to take preventive action to avert damages.

Shaw et al (1996) found that in the case of Taiwan those with higher incomes were less likely to suffer from acute symptoms, although no statistical relationship was found between the incidence of chronic illness and income. Alberini et al (1997) found that the income elasticity of willingness to pay (WTP) to avoid acute illness was 0.33, indicating that environmental quality is not a luxury good.

Employment

The creation of employment may have an impact on the distribution of income within a project area. For example, if miners are made unemployed but workers are employed in other sectors as a result of a mitigation project, this would have a potential influence on the Gini and extended cost-benefit analysis measures.

Ancillary benefits resulting from increased employment may also have distributional considerations. Where those employed come from low-income backgrounds, the degree of inequality in the society would be anticipated to be smaller. The impacts of increased income may have important health benefits, resulting not only from reduced mortality (Markandya, 2000), but also from changes in consumption patterns of fuel sources. Krupnick (1991) highlights the case of indoor air pollution, where a 'substantial' increase in income level results in a change from fuelwood or other highly polluting, inefficient fuel sources for cooking to, for example, electricity or gas.

METHODOLOGICAL FRAMEWORK

In the previous section, a review of the previous work on the estimation of the distributional implications of different elements of mitigation policies or projects was presented. This section presents an explicit framework for the estimation of the distributional impacts of different projects, including the use of distributional weights to measure non-marginal impacts; the use of aggregated household expenditure survey data for the estimation of disaggregated impacts; and the calculation of the distribution of health benefits resultant from reduced electricity production.

Non-marginality of Impacts in the Application of Income Weighting to the Issue of Mitigation

The framework outlined above gives weights to impacts on different income groups. Thus, a dollar of income to a poor person is worth more than a dollar of income to a rich person. However, when one moves away from the issue of marginal to non-marginal impacts on different groups, then difficulties arise in the application of this technique.

The Botswana case study, as outlined in Chapter 7, suggests that in some cases there are reductions in the employment of miners and increases in the employment of other groups. The income lost by the miners and that gained by the other groups has to be considered in the analysis of the project. However, the income weighting approach given in the previous section does not value the loss of earnings as highly as it would a gain in income to the poorer group. Thus, an adjustment needs to be made to the weights to account for non-marginality. The marginal weight defined earlier in equation 6.4 needs to be adjusted. As noted earlier, $\partial W / \partial Y_i$ measures the value of a marginal change in the income of the individual. However, to assess the impacts of a non-marginal change, the weight needed to be applied to the change in income is equal to the median weight. Thus, where Y_i changes more than a marginal unit, an average of the two weights needs to be taken to allow for the changing marginal utility (given that it is not a linear function). Hence, a more generalized system of weights is needed, as set forth in the following equations:

$$\frac{\partial W}{\partial Y_i} = SMU_i = \sum_{n=0}^{j-1} \left[\frac{\bar{Y}}{Y_i + n} \right]^{\varepsilon} / j \text{ for } j>1 \tag{6.6}$$

$$\frac{\partial W}{\partial Y_i} = SMU_i = \left[\frac{\bar{Y}}{Y} \right]^{\varepsilon} \text{ for } j=1 \tag{6.7}$$

$$\frac{\partial W}{\partial Y_i} = SMU_i = \sum_{n=0}^{j+1} \left[\frac{\bar{Y}}{Y_i + n - 1} \right]^{\varepsilon} / |j| \text{ for } j<1 \tag{6.8}$$

where j represents the change in the initial income level. This formulation allows for the inclusion of non-marginal impacts, such as those of employment, in the distributional analysis. This formulation allows a first estimation of the distributional impacts of non-marginal impacts, which would be overestimated using the marginal weighting system outlined above.[11]

Estimation of the Distribution of Control Cost

The 'control cost' of a mitigation option is determined by the nature of the option and the associated level of government or international funding. Where international funding is used to build, for example, a wind farm in Egypt, the control cost falls on the international donor and has no implications for sustainability within the host nation. However, if the wind farm was financed through increases in electricity charges, this has a direct impact on consumers in the host nation and thus the equity issue needs to be assessed. Some mitigation options – for example, the suggestion to raise energy prices – will have such a direct impact. Others may be open to self-funding through the collection of fees (for example, vehicle maintenance checks).

   If the scheme involves revenue raising, such as would be the case with increased electricity charges, techniques similar to those applied for estimating the distributional impacts of environmental taxes can be applied. In the case where a price change affects consumers only, Cornwell and Creedy (1997) suggest the use of household expenditure survey data to estimate the incidence on different income groups, using the following equation:

$$V(q) = \left[\frac{v}{1+v}\right] r_i(q)q \qquad (6.9)$$

where $r_i(q)$ represents the proportion of total expenditure, denoted by q, on the good by group i, and $v/(1+v)$ the proportion of the cost of the good made up by taxes.

Extending the above, as applied in the following analysis, distributional weights can be applied to expenditures on goods and services. The following equation can be used to estimate the weighted costs, drawing on household expenditure survey data:

$$\frac{C_w}{w} = \sum_i \pi_i \beta_i C \qquad (6.10)$$

where C_w is the weighted total cost, β_i is the weight applied to the income or consumption group i, as defined above, π_i represents the proportion of consumption expenditure by group i on the good taxed and C represents the total cost of the project to be financed by the taxation measure. Thus π_i is similar to $r_i(q)q$ in equation 6.9, except that it is represented as a proportion of total consumption expenditure. This analysis assumes that π_i does not vary with the cost of the good. This is perhaps not unreasonable for gaining an estimate of the impacts on given groups of financing a single project, as the impacts on the cost of the good are likely to be marginal.

If the mitigation option would involve increased costs not only to a consumer good but also to inputs into industry, then a more complex evaluation technique involving the use of input–output modelling should be applied. This situation will not be included in the analysis to follow, as it is unlikely to apply in most cases.

Estimation of the Distribution of Ancillary Health Benefits

The inclusion of distributional weights in the integration of distribution into extended cost-benefit analysis is relatively straightforward. However, before outlining the technique, it is necessary to consider the implications of the weighting of health costs. The use of weights in this context is not to suggest that the impacts on poorer groups are more costly than the impacts on richer groups. Rather, it is, as governments have inequality in mind in setting policy, a weight to account for the lower preference that poor groups attach to health damages to themselves. This weight will offset (at least in part) the differences in willingness to pay owing to income levels.

For simplicity, we assume that ancillary health benefits are distributed equally across a given area, with areas closest to power stations most heavily affected. The weighting of such health benefits from the reduced burning of coal to generate electricity can be assessed using the distribution of income relative to population in the affected area. If X_i represents the percentage of the ith income group in the affected area, and Y_i represents the income level for that income group, the distribution weight to be applied to the aggregate health damages is given by W, where:

$$W = \sum_i \frac{X_i}{100} * \left(\frac{\bar{Y}}{Y_i}\right)^{\varepsilon} \qquad (6.11)$$

This is the weight to be applied to a health impact in a given region. If no data on this are available, then an assumption that W is at least equal to 1 may be justified on the basis that there is no difference from the cross-country impacts and thus that there is no impact on inequality. This is a strong assumption, but may be necessary where no data exist on the population affected by pollution.

Estimation of the Distribution of Employment Effects

Changes in the pattern of employment in a region may have two important impacts. First, it directly alters the distribution of income through the wages paid to workers. Second, it reduces or increases the health costs of unemployment, dependent on the impacts on aggregate unemployment. Both these impacts can be valued. The change in income level attributable to the creation (or loss) of a job can be assessed as to its distributional impact through the application of non-marginal weights, as described earlier. The ancillary health benefits or costs can also be weighted using the same methodology.

CASE STUDY OF MITIGATION PROJECTS IN BOTSWANA

This section builds on the case study described in Chapter 7 which assesses the inclusion of ancillary benefits in the assessment of mitigation projects in Botswana. The impact of the inclusion of distribution into the analysis of projects is assessed using distributional weights.

This study does not attempt to analyse the distribution of the primary benefits of mitigation – ie, the benefits resultant from reduced climate change as a result of mitigation. The benefits from reduced climate change are widely acknowledged to be difficult to assess, given the uncertainties involved and a lack of consensus in the literature on the costing of such impacts.

In the preceding sections many issues relating to the measurement of distributional effects have been discussed. Different measurement techniques and problems with these have been addressed, along with a framework for the identification of distributional impacts. In this case study, the distributional weights approach is applied to a concrete example for the case of Botswana. Distributional weights allow for the identification of the broad measures of the distributional impacts of policies. They also allow for easy comparison with a no-weights case. As has been discussed, distributional weighting is by no means a perfect technique, and the arguments against such weighting, including those described earlier, need to be considered. However, the need for a technique to assess distribution is clear, and while arguments exist with regard to its appropriate use, the other techniques on offer are either data hungry or subject to similar problems. Thus, in the absence of a better technique, the case study presented here uses the weighting technique.

There remains the need for a close examination of the impacts on vulnerable groups through stakeholder analysis, as suggested earlier in this chapter. However, due to limited data in the Botswana context, this is something that would require further research. The study below considers only the impact on different groups within Botswana by income and consumption group and does not attempt to evaluate the relative political importance of groups or industries.

Overview

The case study for Botswana described in Chapter 7 considers seven projects: road paving; efficient lighting; power factor correction; construction of a pipeline for fuel oil; the establishment of a central PV plant; improved industrial boilers; and vehicle inspection. These seven projects will be assessed with regard to their implications on different income groups and their impact on the overall equity situation in Botswana.

Botswana has an unequal income distribution. The Household Income and Expenditure Survey (HIES) 1993–1994 indicates that the Gini coefficient for disposable cash income is 0.638. The Gini measure falls significantly when in-kind income is taken into consideration to 0.537, which, however, is still unequal relative to other developing nations.[12]

Costs and Ancillary Benefits to be Examined

Table 6.2 presents a taxonomy of the costs and ancillary benefits that will be examined in determining the distributional aspects of mitigation strategies in Botswana. The ancillary benefits resulting from increased employment and improved health as a result of a vehicle maintenance project are assumed to have positive distributional implications. Poorer sections of the community are more likely to live close to roads, as the poor are often assumed to be less sensitive to environmental degradation than the rich.[13] On the other hand, reduced employment is assumed to have a negative distributional impact. The distributional nature of the costs of such a project depends on the method used to finance the project.

This case study will attempt to assess the costs and benefits of climate mitigation projects, drawing on the methodologies of income weighting and the impacts on different key groups in society presented earlier in this chapter.

The distributional implications of changes in employment is based on data linking income from employment to overall incomes of different groups (Zhou, personal communication). The impacts on consumer spending will be assessed drawing on the Botswana HIES (CSO, 1995). The distribution of health impacts are evaluated using data derived from the Botswana census. The implications of different tax policies to fund or partially fund projects will also be examined, drawing on the HIES data.

In addition to the costs and benefits identified in the table, other smaller secondary impacts may be experienced. For example, road paving and vehicle inspections may have secondary impacts on the maintenance costs of vehicles. Road paving may reduce the number of repairs needed due to punctures, leading to impacts on employment in vehicle maintenance. Vehicle inspection may lead

Table 6.2 *Distributional implications of costs and ancillary benefits of mitigation projects*

Strategy	Cost/ancillary benefit considered	Expected distributional implication	Justification
Road pavement	Cost	?	Depends on project financing
	Reduced health costs from lower $SO_2/NO_x/PM$ emissions	+	Poorer sections of the community likely to be closer to roads (hedonic pricing) and unable to make defensive expenditures
	Increased employment	+	Construction workers likely to be unskilled
Efficient lighting	Cost	?	Depends on project financing
	Reduced health costs from lower $SO_2/NO_x/PM$ emissions	+	Poorer sectors of the community likely to live closer to power generators and unable to finance defensive expenditures
	Reduced employment in mines	–	Increased inequality as miners revert to subsistence or lower paid jobs
Power factor correction	Cost	?	Depends on project financing
	Reduced health costs from lower $SO_2/NO_x/PM$ emissions	+	Poorer sectors of the community likely to live closer to power generators and unable to finance defensive expenditures
	Reduced employment in mines	–	Increased inequality as miners revert to subsistence or lower paid jobs
Pipeline construction	Cost	?	Depends on project financing
	Reduced health costs from lower $SO_2/NO_x/PM$ emissions	+	Poorer sectors of the community likely to live closer to power generators and unable to finance defensive expenditures
	Reduced employment in transport sector	–	Increased inequality as transport workers revert to subsistence or lower paid jobs
	Increased employment in construction phase	+	Short-term increase in employment for the low-skilled
Central PV plant	Cost	?	Depends on project financing
	Reduced health costs from lower $SO_2/NO_x/PM$ emissions	+	Poorer sectors of the community likely to live closer to power generators and unable to finance defensive expenditures
	Reduced employment in mines	–	Increased inequality as miners revert to subsistence or lower paid jobs
	Construction of PV plant	+	Short-term increase in employment for the low-skilled
Industrial boilers	Cost	?	Depends on project financing
	Reduced health costs from lower $SO_2/NO_x/PM$ emissions	+	Poorer sectors of the community likely to live closer to power generators and unable to finance defensive expenditures
	Reduced employment in mines	–	Increased inequality as miners revert to subsistence or lower paid jobs
Vehicle inspection	Cost	? (likely +)	Depends on project financing
	Reduced health costs from lower $SO_2/NO_x/PM$ emissions	+	Poorer sections of the community likely to be closer to roads (hedonic pricing) and unable to make defensive expenditures

to early identification of problems or to increased maintenance due to abuse by mechanics. These secondary impacts are likely to be small, but, as we noted earlier, they could be important in some cases. Only public consultation will reveal which impacts need further analysis. This study has not been able to make such an assessment and to that extent the analysis here is incomplete.

Distributional Analysis

This section will employ the framework for distributional weighting and the impacts of the different projects on the key interest groups in Botswana.

Estimation of Income Weights

The income weights that may be used in the Botswanan case are based around the average Botswana cash income of pula 831 and are shown in Table 6.3. These weights are calculated using inequality parameters of 1, 1.5 and 2 to give an indication of the effects of the different levels of importance attributed to the issue of inequality. Thus for Botswana, income weights are derived using the following formula:

$$W = \left(\frac{\bar{Y}}{Y_i}\right)^{\varepsilon} \Rightarrow (831/yi)^{\varepsilon} \qquad (6.12)$$

Table 6.3 *Income weights for Botswana*

Income level Monthly (pula)	Value of inequity parameter		
	1.0	1.5	2.0
100	8.3	24.046	69.406
200	4.166	8.502	17.351
300	2.777	4.628	7.712
400	2.083	3.006	4.338
500	1.666	2.151	2.776
600	1.389	1.636	1.928
700	1.190	1.298	1.416
800	1.041	1.063	1.084
900	0.926	0.891	0.857
1000	0.833	0.760	0.694
1100	0.757	0.659	0.574
1200	0.694	0.578	0.482
1300	0.640	0.513	0.411
1400	0.595	0.459	0.354
1500	0.555	0.414	0.308
1600	0.521	0.376	0.271
1700	0.490	0.343	0.240
1800	0.463	0.315	0.214
1900	0.438	0.290	0.192
2000	0.417	0.269	0.174

Source: Author's own calculations based on CSO (1995) and methodology as described in text

This indicates that the marginal value of an extra pula to an individual earning 200 pula is P4.16, P8.50 or P17.46, depending on the inequality weight applied. The determinants of the inequality aversion parameter are described earlier in this chapter.

Estimation of Consumption Weights

For the purpose of comparison, an evaluation of the impact of using consumption rather than income weights was conducted. It has been proposed that consumption is a better proxy for the assessment of the distributional impacts of different cost increases (Poterba, 1989), as individuals distribute consumption across their lifetime to compensate for uneven income flows. The application of consumption weights for the Botswana case was particularly interesting as the Botswana HIES suggested that the lowest income group consumed more on average than the next highest income group. The weights applied are shown in Table 6.4 below, for selected income groups. These weights were assessed according to a mean consumption level of 716 pula.

Table 6.4 *Botswana consumption weights*

				Income level (pula/month)					
	59.49	150.74	246.74	398.1	713.76	1215.47	1725.81	245.2	6376.53
Inequality aversion				*Consumption expenditure (pula/month)*					
ε	166.98	142.51	222.29	303.52	465.53	845.59	1139.15	1820.44	3685.85
1.0	4.29	5.03	3.22	2.36	1.54	0.85	0.63	0.39	0.19
1.5	8.88	11.26	5.78	3.62	1.91	0.78	0.50	0.25	0.09
2.0	18.39	25.25	10.38	5.57	2.37	0.72	0.40	0.15	0.04

Source: Author's calculations based on Botswana HIES (1993–1994)

Control Costs

The distribution of the costs of the projects relates, as mentioned earlier in this chapter, to the different measures used to fund projects. The distributional impacts of the different measures for raising revenue was estimated using a simple incidence analysis, based on data from the Botswana HIES, 1993–1994. The adjustment factors applied for this analysis are shown in Table 6.5.

These weights were calculated using the following equation:

$$W = \sum_i \pi_i \beta_i = \sum_i \left(\frac{\bar{Y}}{Y_i}\right)^{\varepsilon} \bullet \left(\frac{x_i}{X}\right)$$

(6.13)

This equation makes it possible to operationalize equation 6.10 above. On the right-hand side, $\bar{Y}$ represents average income/consumption, Y_i the income or consumption of group i, x_i the consumption of good x by income/consumption group i and X the total consumption of good i. These weights can then be multiplied by the project costs to facilitate the application of weighted cost-benefit analysis in the case of climate mitigation policy. The lower the value of W the less regressive is a tax based on good X.

Table 6.5 *Adjustment factors based on income weights for different commodities in Botswana*

Weight for	Inequality aversion parameter		
	1.0	1.5	2.0
Car ownership	0.89	1.26	2.08
Petrol consumption	0.62	0.66	0.84
Taxation revenues (1)	0.73	0.93	1.43
Electric lighting (2)	1.57	2.44	4.12
Electricity consumption	0.47	0.43	0.47

Notes:
(1) Taxation weights based on differences between gross and disposable income
(2) Electric lighting: weight based on consumption of lightbulbs

It can be seen from Table 6.5 that measures which increase the cost of electricity consumption will be less regressive than raising tax in line with prior tax policy. The ownership of cars and electric lighting is more evenly distributed across the income frontier than petrol expenditures, indicating that an increase in the cost of fuel would be less regressive than a tax on car ownership. The underlying calculations for these weights are presented in Appendix 6.1.

The following analysis assumes that a project may be self-funded, by raising charges on the GHG-related good or funded through general taxation. Options that may be considered to have more potential for cost recovery through charges include the efficient lighting scheme and the vehicle inspection scheme. Both of these options have considerable potential for cost recovery through direct charging on lightbulbs and inspections. However, all projects were assessed regarding to the distributional impact of raising a charge on the consumption of the good involved, be it petrol, cars or electricity. The assumptions made in this analysis are presented in Appendix 6.1.

Results and Implications for Project Selection

Results Using Income Weights

The results of the project analysis using income weights, based on a mid-point inequality aversion parameter of 1.5 and a discount rate of 3 per cent, are given in Table 6.6.

The figures in this table reflect measures of the net social costs per tonne of CO_2, with the no adjustment case giving a cost estimate in which equity considerations are ignored. The weighted case gives some indication of the equity impact by allocating a different value to impacts on individuals with low income. The tables in this section should be treated with care – they are of use in formulating policy, but the weights applied to different impacts are open to some discussion, as mentioned above. As can be seen from the table, there are significant impacts on the net costs per tonne for some projects, while other projects are largely unaffected. This reflects the different distributional characteristics of the control cost and ancillary benefits. For example, efficient lighting costs decrease dramatically, while the paved roads scheme is only slightly impacted by the application of adjustment factors.

Table 6.6 *Project analysis with income weights*

Project name	No equity adjustment	With equity adjustment Self-funding	Taxation
	$\varepsilon = 1.5$, $r = 3\%$ US$/t$CO_2$		
Vehicle inspection	1.18	6.53	1.21
Paved roads	−180.08	−160.34	−151.71
Efficient boilers	−224.87	na	−235.72
Efficient lighting	−194.28	−779.67	−1016.32
Central PV electricity	−209.74	−229.37	−156.81
Petroleum pipeline	−76.38	−14.21	94.15
Power factor correction	−265.79	−250.69	−237.33

In terms of the ranking of projects, the application of income weights has some effect. The rankings of the different projects under the assumptions above are shown in Table 6.7 below. These results indicate that the consideration of equity is important in the framing of climate mitigation policy. Schemes that have higher net costs in the absence of equity considerations may have lower net costs if these costs fall on high-income groups. Schemes where the costs fall on poorer groups are considered less attractive under this project evaluation method. The impact of the consideration of equity in the Botswana case is most notable for the efficient lighting scheme, which rises from fourth best project to best project in terms of the cost per ton of CO_2, indicating that if a policy-maker were to wish to meet equity objectives, taking into account the other social impacts of the projects, then efficient lighting should be encouraged.

Results Using Consumption Weights

The implications of the inclusion of distributional weighting based on consumption are shown in Tables 6.8 and 6.9. It can be seen that the application of such

Table 6.7 *Rankings of projects with income weights*

Project name	No equity adjustment	With equity adjustment Self-funding	Taxation
	$\varepsilon = 1.5$, $r = 3\%$ Ranking		
Vehicle inspection	1	6	6
Paved roads	5	4	5
Efficient boilers	2	na	3
Efficient lighting	4	1	1
Central PV electricity	3	3	4
Petroleum pipeline	6	5	7
Power factor correction	1	2	2

weights has a significant impact on some projects – notably, efficient lighting and the petroleum pipeline projects. The petroleum pipeline project, if funded through raising petrol prices, would have been more regressive in nature, owing to the consumption patterns of petrol.

Table 6.8 *Project analysis with distributional weights on consumption*

Project name	$\varepsilon = 1.5$, $r = 3\%$ US$/tCO_2		
	No equity adjustment	With equity adjustment	
		Self-funding	Taxation
Vehicle inspection	1.18	1.65	1.21
Paved roads	−180.08	−151.58	−149.12
Efficient boilers	−224.87	na	−229.52
Efficient lighting	−194.28	−371.93	−459.46
Central PV electricity	−209.74	−229.61	−197.27
Petroleum pipeline	−76.38	−20.49	10.37
Power factor correction	−265.79	−250.73	−244.78

In terms of the ranking of projects, Table 6.9 shows that there are significant impacts of the inclusion of distributional weights on consumption. Notably, efficient lighting goes from being the fourth ranked project to the highest ranked project. The petroleum pipeline project also falls down the rankings when the control costs are weighted according to general taxation. Overall, vehicle inspection performs most badly.

Comparison: Income Weights and Consumption Weights

The above results show that there are significant impacts to be found on project selection as a result of the application of income or consumption weights. Projects that may have a significant positive impact on the poorer groups in society are

Table 6.9 *Rankings with distributional weights on consumption*

Project name	$\varepsilon = 1.5$, $r = 3\%$ Ranking		
	No equity adjustment	With equity adjustment	
		Self-funding	Taxation
Vehicle inspection	7	6	6
Paved roads	5	4	5
Efficient boilers	2	na	3
Efficient lighting	4	1	1
Central PV electricity	3	3	4
Petroleum pipeline	6	5	7
Power factor correction	1	2	2

ranked more highly as a result of the application of weights, while those which have a detrimental impact can be identified. In the above case study, efficient lighting clearly has significant distributional implications and thus is ranked more highly when weights are used.

The results above show that the application of income weights leads to stronger impacts on the net costs per ton of carbon dioxide abated than consumption weights. This is to be expected, as consumption is likely to be more evenly distributed than income. For example, the efficient lighting project has a net cost of -US\$194 per tonne of CO_2 when equity is not taken into consideration, yet this changes to a range of -US\$780 to -US\$1016 per tonne depending on the funding source under income weights. The equity aspect is less dramatic under the consumption weighting, with the same net costs being in the order of -US\$372 to -US\$459 per tonne of CO_2 abated.

In the above analysis no impact was detected on the relative ranking of the projects. This could be due to the fact that some ancillary benefits could not be assigned weights due to the lack of data. However, it is also broadly indicative that consumption and income weights, though of different magnitudes, lead to broadly the same results in terms of the selection of different projects. Consumption measures are often more difficult to obtain than income measures, and that income acts as a broad proxy for consumption is shown above.

The issue of whether income or consumption weights should be applied is open to debate. This hinges on whether equality in income or consumption is the target of policy. Sen (1992) discusses the problem in the analysis of inequality – the crucial problem of 'equality of what?' being central to the debate. The above analysis shows that, for the climate mitigation policy case outlined, although the absolute values of the different projects differ as a result of the use of income or consumption weights, their rankings do not. This need not necessarily be the case, however, as the inequality aversion parameter may also have an impact and it is to this that we now turn.

Sensitivity Analysis: Changing the Inequality Aversion Parameter

The inequality aversion factor, representing the weight given to reducing inequality by a policy-maker, may be crucial in determining the relative strengths of different projects. Taking values of 1 and 2 for this parameter, a sensitivity analysis was performed to check the impact of varying it on project choice in the case of both income and consumption weighting.

Income weighting

The impacts of changing the inequality aversion parameter in the case where income weights are applied are shown in Table 6.10 for the case where projects are self-financing. As can be seen from the table, the rankings are fairly stable, although the absolute values may change substantially. The only projects to change rankings are vehicle inspection and the petroleum pipeline projects.

The sensitivity of the analysis of the projects in the case where projects are financed from general taxation is shown in Table 6.11. As can be seen from the table, when the inequality aversion parameter is set at 2 (ie, the policy-maker is very concerned about inequality), the rankings of projects change more than under self-financing.

Table 6.10 *Sensitivity analysis: changing the inequality aversion parameter under income weights when projects are self-financing*

Project name	$\varepsilon=1$	% Change	Ranking	Change	$\varepsilon=1.5$	Ranking	$\varepsilon=2$	Ranking	Change
Vehicle inspection	2.04	−68.72	6	0	6.53	6	24.38	5	1
Paved roads	−142.26	−11.28	4	0	−160.34	4	−221.45	4	0
Efficient boilers	na				na		na		
Efficient lighting	−351.65	−54.90	1	0	−779.67	1	−2319.27	1	0
Central PV electricity	−223.75	−2.45	3	0	−229.37	3	−255.99	3	0
Petroleum pipeline	−24.67	73.61	5	0	−14.21	5	25.04	6	1
Power factor correction	−249.65	−0.41	3	0	−250.69	2	−250.06	2	0

Table 6.11 *Sensitivity analysis: changing the inequality parameter under income weighting when projects are funded by general taxation*

Project name	$\varepsilon=1$	Ranking	Change	$\varepsilon=1.5$	Ranking	$\varepsilon=2$	Ranking	Change
Vehicle inspection	1.00	6	0	1.21	6	7.26	6	0
Paved roads	−139.96	5	0	−151.71	5	−189.92	4	1
Efficient boilers	−229.04	3	0	−235.72	3	−259.99	2	1
Efficient lighting	−432.96	1	0	−1016.32	1	−3112.93	1	0
Central PV electricity	−200.43	4	0	−156.81	4	1.74	5	−1
Petroleum pipeline	4.20	7	0	94.15	7	421.04	7	0
Power factor correction	−245.36	2	0	−237.33	2	−208.16	3	−1

Consumption Weighting

In the case where projects may be self-financed, Table 6.12 shows that although absolute values of the net project costs per tonne of CO_2 change dramatically in some cases, the relative rankings of the projects do not change at all. This shows that the values may be sensitive, but the impact of using different inequality parameters on the selection between projects is small in the above case. Thus, if the government were committed to mitigation measures and to equity objectives in the above case, the selection of the inequality aversion parameters would not be important.

Table 6.12 *Sensitivity analysis: Changing the inequality aversion parameter under consumption weights when projects are self-financing*

Project name	ε=1	Ranking	Change	ε=1.5	Ranking	ε=2	Ranking	Change
Vehicle inspection	1.04	6	0	1.65	6	3.05	6	0
Paved roads	−142.44	4	0	−151.58	4	−183.23	4	0
Efficient boilers	na	na	na	na	na	na	na	na
Efficient lighting	−264.44	1	0	−371.93	1	−578.39	1	0
Central PV electricity	−222.60	3	0	−229.61	3	−228.53	3	0
Petroleum pipeline	−24.54	5	0	−20.49	5	−2.37	5	0
Power factor correction	−249.44	2	0	−250.73	2	−250.53	2	0

If projects were financed by increasing general taxation, the results are somewhat different. Table 6.13 shows the main results. In the case where the inequality aversion parameter is reduced to 1, the petroleum pipeline project becomes slightly more favourable compared to vehicle inspection. When the parameter rises from 1.5 to 2, the impact is to make the paved roads scheme more favourable, with cost per tonne of CO_2 declining from −US$149 to -US$178. This moves the paved roads scheme above central PV electricity for which cost per tonne CO_2 increases from −US$197 to US$170.

Overall, although the absolute values in terms of the cost per tonne of CO_2 can be sensitive to changes in the inequality aversion parameter, the rankings do not vary greatly, with the choice of income or consumption weights not being critical to this conclusion. This suggests that the findings are fairly robust. Thus, policy-makers concerned about the secondary impacts of climate mitigation projects in terms of the distributional impacts of such projects can determine those projects that are less distributionally harmful and the least harmful ways of raising revenues to finance such projects. This is true for both income weighting and consumption weighting, though income weighting becomes more unstable as the inequality aversion parameter rises.

Summary

The inclusion of distributional weights has significant impacts on the relative rankings of different projects. Significant changes in the net costs per tonne of CO_2

Table 6.13 *Sensitivity analysis: changing the inequality parameter under consumption weighting when projects are funded by general taxation*

Project name	$\varepsilon=1$	Ranking	Change	$\varepsilon=1.5$	Ranking	$\varepsilon=2$	Ranking	Change
Vehicle inspection	0.75	7	1	1.21	6	1.94	6	0
Paved roads	−141.49	5	0	−149.12	5	−177.74	4	1
Efficient boilers	−227.81	3	0	−229.52	3	−233.77	4	1
Efficient lighting	−313.23	1	0	−459.46	1	−733.79	1	0
Central PV electricity	−208.49	4	0	−197.27	4	−169.52	5	−1
Petroleum pipeline	−12.65	6	1	10.37	7	66.48	7	0
Power factor correction	−246.85	2	0	−244.78	2	−239.67	2	0

abated can be seen from the above results, with the changes being most significant where weights based on income rather than consumption are used. Significant impacts can also be seen in the relative rankings of different projects, with those with the most distributionally negative cost distribution being identified through the use of the weights.

Some critics of distributional weights have argued that the inequality aversion parameter assigned is arbitrary. While this may be the case to some extent, some studies have resulted in empirical estimates for this parameter and the above analysis shows that while the overall results in terms of project selection are sensitive to some equity adjustment, they are not that sensitive to the actual value of the inequality aversion parameter when the latter takes a range of plausible values. However, best practice would suggest that sensitivity analysis of the type shown above should always be conducted when applying the weighting system.

Many of the estimates of the net costs per tonne of CO_2 abated shown above were negative, indicating a net gain to society of undertaking the mitigation option. These figures may be underestimates of the net gain, as no weights have been assigned to health benefits, which, as discussed earlier, may be expected to have positive distributional outcomes as the poor are often most exposed to pollution.

CONCLUSIONS

Equity considerations are of importance in determining the sustainability and potential for implementation of climate policy. Intergenerational equity concerns highlight the necessity for action, but once action has been deemed necessary intragenerational concerns are of greater importance. With increased focus on issues such as poverty alleviation and other development actions as co-benefits of climate mitigation strategy, the determination of the distributional implications of different policy measures takes on a greater degree of importance. The importance of the distributional impacts are not only based on the desire for policy to be broadly ethical, but also as the distribution may hamper the successful implementation of these measures.

Distributional weighting has been suggested in the literature as a technique for the inclusion of distributional concerns in cost-benefit analysis. The determination of the weights to be applied is important and should be determined from the priorities assigned to the equity issue by policy-makers. This chapter does not claim that this technique is without flaw, but the use of weights does allow easy comparison of projects and their impacts. Other possible measures of the distributional impact of projects include the Gini measure and direct measures of impacts on poverty. Of particular importance to the successful implementation of policy is that the impact on key stakeholders is identified. This chapter presents one possible technique for this, using a distributional matrix.

Different techniques exist for the measurement of the distributional impact of policies. These include partial equilibrium models and input–output models. The former may be applied where there is no significant impact on producer prices. The latter allow more extensive examination of the impacts of taxes across the economy.

Different measures of control cost and ancillary benefits add complexity to the selection of policy. The distributional nature of some ancillary benefits have not been assessed, and where they have the evidence is mixed. For the distribution of control cost, data from household expenditure surveys may be used to assess the impacts on different income groups of a tax on the commodity in question.

The case study in of Botswana illustrates how distributional weighting of the control cost and ancillary benefits of climate mitigation projects may impact project analysis. As knowledge of the distribution of ancillary benefits expands, such weighting may yet further impact on the analysis of projects.

In terms of the inequality aversion parameter used to weight the cost of a project, the findings of this case study are interesting. Overall, although the absolute values in terms of cost per tonne of CO_2 can be sensitive to changes in the inequality aversion parameter, the rankings do not vary greatly. This suggests that the findings are fairly robust, which mitigates the argument somewhat that weighting is arbitrary and may lead to an incorrect project selection. Where inequality is considered a problem, the weights applied must be subjected to sensitivity analysis to confirm whether the weights are having a disproportionate impact on the selection of projects.

Equity concerns are important in terms of project selection for both national governments and to the wider international community. This concern has been enhanced by the inclusion of sustainable development requirements within the Kyoto Protocol. Academic debate on the nature of inequality will continue, but for project selection not to consider reduction of income inequality and poverty would be fallacious, as it is an important issue, particularly in many developing countries.

The main impacts on mitigation policy that this chapter shows are as follows:

- Projects should be carefully assessed regarding the nature of the impacts, both direct and ancillary, that result.
- The method of the funding of projects, including methods for cost recovery through charges on related goods and services, may have important implications for the distributional impacts. Where the cost falls mainly on poorer groups, this may mean the government may prefer to use either outside funding or general taxation to cover the costs of the project. The latter is likely to be particularly true in developed countries where access to Clean Development Mechanism (CDM) or joint implementation (JI) funds are not available.[14]
- Assessments of the distribution of ancillary health impacts are needed to give a more complete picture of the distributional impact of mitigation policies that reduce the levels of local air pollutants, such as those that reduce fossil fuel use. It is likely, however, that mitigation policy will have significant positive impacts on the health of poor communities, which are often most exposed to high levels of pollution.
- Many of the projects identified above have net costs per tonne of CO_2 below zero, indicating a social gain from mitigation activity, even when distributional aspects are taken into account. There may, however, be significant barriers to implementation, including social and institutional barriers mentioned in Chapter 2. For such projects, these need to be examined, possibly using stakeholder analysis.

The equity implications of climate mitigation projects must be considered alongside the broader impacts of such projects. The impact on social capital, on the health and well-being of individuals, must be evaluated in a consistent framework with the distributional implications of climate policy. For an assessment of the full impact on sustainable development, more work is obviously needed. However, this chapter represents an important step towards the consideration of equity in the assessment of projects and presents different techniques that may be applied.

Although the assessment of the distributional consequences of mitigation options is in its infancy, its inclusion is important to provide guidance on the broad nature of the impact for policy purposes at national and international level. In the design of the CDM, discussed in Chapter 8, impacts on intragenerational equity need to be considered alongside those of social capital in order that sustainable development objectives can be met. If equity is ignored, serious impacts on the poor may result or mitigation may be restricted by key interest groups applying pressure on national governments. Thus it is important to assess the effects of mitigation options on both different income groups and key stakeholders, so as not to hamper the implementation of the Kyoto Protocol and moves towards the promotion of sustainable development in the developing world.

NOTES

1   My thanks go to the Danish Research Council for financial support in the writing of this chapter, and to Anil Markandya of the University of Bath for guidance and comments. Thanks are also due to the UNEP Collaborating Centre on Energy and Environment for the use of facilities and to Kirsten Halsnaes, Anne Olhoff and Kim Olsen for comments on the draft which have greatly improved the structure and scope of this chapter. All errors are, of course, my own.

2   Meier (1995), p22. This is based on a cross-country comparison of growth and inequality. Meier notes that this runs against what may be felt to be intuition – ie, that higher levels of inequality promote growth through savings and capital accumulation. Meier cites Williamson (1991) in suggesting that there is historical proof that rising inequality made capital accumulation more different rather than less in the 19th century.

3   The prescriptive approach leads to estimates of the social rate of time preference, which some advocate should be used as the discount rate. This can be written as follows: $SRTP = \rho + \theta$, where ρ shows the discounting of future generations' utility and θg represents the discounting needed to account for rising consumption. The growth rate of per capita consumption is taken to be g. The value of ρ reflects the level of impatience or myopia. θg is used to take account of diminishing marginal utility of consumption. For a detailed description of the social rate of time preference see IPCC (1996).

4   Cropper and Laibson (1999) state three preconditions for differential discounting: that there is separability in production, separability in consumption and that there is government control over the environmental production, with a time lag for changes.

5   Recent efforts to quantify the impacts of inequality on development include Stymne and Jackson (2000), who calculated measures of the impact of inequality on sustainable welfare. Stymne and Jackson found that the welfare loss due to inequality ranged from 5 to 10 per cent of GDP in Sweden and 6 to 14 per cent of gross domestic product (GDP) in the UK, based around an inequality parameter of 0.8.

6   For a recent review of the impacts of poverty on the environment, see Markandya (2001).

7   Little and Mirrlees (1974) proposed the use of consumption, rather than income, weights. While this may allow more accurately for the consideration of lifetime income and savings to enter the analysis, data on income is more readily available than that on consumption.

8   In developing countries the impact on house prices of environmental quality may not be that large, given that willingness to pay for environmental improvement in poor areas has been shown to be low. Thus, in the analysis that follows the distributional impact of mitigation programmes on health but not on property price are considered.

9   The problem with this approach is, of course, the implicit assumption that past policy decisions have been taken in a rational manner. To the extent this is not true, the resulting estimates are flawed.

10   Barriers to implementation may consist of social barriers, such as those highlighted here and in Chapter 4, or institutional barriers. An overview of such barriers is presented in Chapter 2, and the issue is further discussed in Chapter 5 of IPCC (2001).

11   Little and Mirrlees (1974) proposed a similar formulation, recommending regarding the utility difference 'as the sum of a large number of small contributions to social value, arising from small changes in consumption which gradually change [the consumption level], each of these small changes being multiplied by the appropriate weighting factor'.

12   Scholars often suggest that 0.4 for the Gini measure is a critical value – numbers higher than that sustained for a long time can result in social unrest and disintegration.

13   For a review of the linkages between environmental degradation and the poor see Markandya (2001).

14   For a wider discussion of the impacts of CDM on sustainable development, including case studies for Brazil and India, see Chapter 8.

15   Zhou, personal communication.

REFERENCES

ADB (2001) *Guidelines for the Economic Analysis of Projects*, ADB, Manila, http://www. adb.org/documents/guidelines/eco_analysis/introduction.asp

Alberini, A, Cropper, M, Fu, T, Krupnick, A, Liu, J, Shaw, D and Harrington, W (1997) 'Valuing Health Effects of Air Pollution in Developing Countries: The Case of Taiwan,' *Journal of Environmental Economics and Management*, Vol 34, pp107–126

Atkinson, A B (1970) 'On the measurement of inequality', *Journal of Economic Theory*, No 22, pp44–263.

Barbier, E, Markandya, A and Pearce, D (1990) 'Environmental sustainability and cost-benefit analysis', *Environment and Planning A*, No 22, pp1259–1266

Blundell, R, Brewning, M and Meghit, C (1994) 'Consumer demand and the life-cycle allocation of household expenditures', *Review of Economic Studies*, No 61, pp57–80, cited in Stymne and Jackson (2000).

Bonguignon, F, de Melo, J and Morrisson, C (1991) 'Poverty and Income Distribution during Adjustment: Issues and Evidence from the OECD Project', World Bank Country Economics Department, working paper 810

BPC (1996) Annual Report, Botswana Power Corporation, Botswana, cited in Zhou (1996).

Brent, R J (1990) *Project Appraisal for Developing Countries*, New York University Press, New York

Brooks, N and Sethi, R (1997) 'The Distribution of Pollution: Community Characteristics and Exposure to Air Toxics', *Journal of Environmental Economics and Management*, Vol 32, No 2, pp232–250

Christiansen, V and Jensen, E S (1978) 'Implicit Social Preferences in the Norwegian System of Social Preferences', *Journal of Public Economics*, Vol 10, pp217–245

Cline, W (1993) 'Give Greenhouse Abatement a Fair Chance', *Finance & Development*, Vol 30, No 1, pp3–5

Cline, W (1999) 'Discounting for the Very Long Term', in P Portney and J Weyant (eds) *Discounting and Intergenerational Equity*, Resources for the Future, Washington, DC

Cornwell, A and Creedy, J (1997) *Environmental Taxes and Economic Welfare: Reducing Carbon Dioxide Emissions*, Edward Elgar, Cheltenham

Cropper, M and Laibson, D (1999) 'The Implications of Hyperbolic Discounting for Project Evaluation', in P Portney and J Weyant (eds) *Discounting and Intergenerational Equity*. Resources for the Future, Washington, DC

CSO (1995) *Household Income and Expenditure Survey*, Central Statistics Office, Botswana

Dixit, A (1996) *The Making of Economic Policy: A Transaction-Cost Politics Perspective*, The MIT Press, Cambridge, MA

Eskeland, G and Kong, C (1998) 'Protecting the Environment and the Poor: A Public Goods Framework, and an Application to Indonesia', Working Paper 1961, Development Research Group, The World Bank, Washington, DC

Feldstein, M (1998) 'Income Inequality and Poverty', NBER working paper no W6770, National Bureau of Economic Research, Cambridge, MA

Frisch, R (1959) 'A Complete Scheme for Computing all Direct and Cross Elasticities in a Model with Many Sectors', *Econometrica*, Vol 27, pp177–196

Gini, C (1912) *Variabilita e Mutabilita*, Bologna

IPCC (1996) *Climate Change 1995: Economic and Social Dimensions of Climate Change*, Cambridge University Press, Cambridge

IPCC (2001) *Climate Change 2001: Mitigation*, Cambridge University Press, Cambridge

Kopp, R and Portney, P (1999) 'Mock Referenda for Intergenerational Decisionmaking', in P Portney and J Weyant (eds) *Discounting and Intergenerational Equity*, Resources for the Future, Washington, DC

Krupnick, A (1991) 'Urban Air Pollution in Developing Countries: Problems and Policies', Discussion Paper QE91–14, Resources for the Future, Washington, DC

Lind, R C (1995) 'Intergenerational Equity, Discounting, and the Role of Cost-Benefit Analysis in Evaluating Global Climate Policy', *Energy Policy*, Vol 23, No 4/5, pp379–389, reprinted as Chapter 27 in Tietenberg (1997) *The Economics of Global Warming*, Edward Elgar, Cheltenham

Lind, R C and Schuler, R (1998) 'Equity and Discounting in Climate-Change Decisions', in W Nordhaus (ed) *Economics and Policy Issues in Climate Change*, Resources for the Future, Washington, DC

Little, I M D, and Mirrlees, J A (1974) *Project appraisal and planning for developing countries*, Heinemann Educational Books, London

Markandya, A (1998) 'The Indirect Costs and Benefits of Greenhouse Gas Limitations'. Handbook Reports, UNEP Collaborating Centre on Energy and Environment, Roskilde, Denmark

Markandya, A (2000) 'Employment and Environmental Protection', *Environmental and Resource Economics*, Vol 15, No 4, pp297–322

Markandya, A (2001) 'Poverty, Environment and Development' in H, Folmer, H L Gable, S Gerking and A Rose (eds) (2001) *Frontiers of Environmental Economics*, Edward Elgar, Cheltenham

Meier, G M (1995) *Leading Issues in Economic Development*, Oxford University Press, Oxford

Metcalf, G (1998) 'A Distributional Analysis of an Environmental Tax Shift', NBER working paper No W6546, National Bureau of Economic Research, Cambridge, MA

Murty, MN et al (1992) 'National Parameters for Investment Project Appraisal in India', working paper no E/153/92, Institute of Economic Growth, University Enclave, Delhi

Murty, MN and Markandya, A (2000) *Cleaning Up the Ganges: A Cost Benefit Analysis of the Ganga Action Plan*, Oxford University Press, Oxford

ODA (1995) *Guidance Notes on how to do Stakeholder Analysis of Aid Projects and Programmes*, Social Development Department, ODA (now DFID), London

Poterba, J (1989) 'Lifetime Incidence and the Distributional Burden of Excise Taxes', AEA Papers and Proceedings, May 1989

Rawls, J (1971) *A Theory of Justice*, Belknap Press, Harvard University Press, Cambridge, MA

Schelling, T C (1992) 'Some Economics of Global Warming', *American Economic Review*, Vol 82, No 1, pp1–14, reprinted as Chapter 2 in Tietenberg, T (1997) *The Economics of Global Warming*, Edward Elgar, Cheltenham

Schwartz, J and Winship, C (1979) 'The welfare approach to measuring inequality' in Schuessler, K (ed) (1980) *Sociological Methodology*, Jossey-Bass Publishers, San Francisco

Scott, M F G, J D Macarthur and D M G Newbury (1976) *Project Appraisal in Practice: The Little/Mirrlees Method Applied to Kenya*, Heinemann, New York

Sen, A (1992) *Inequality Re-examined*, Clarendon Press, Oxford

Sen, A (1997) *On Economic Inequality*. Clarendon Press, Oxford

Shah, A and Larsen, B (1992) 'Carbon Taxes, the Greenhouse Effect, and Developing Countries', World Bank Policy Research Working Paper Series No 957, The World Bank, Washington, DC

Shaw, D, Fu, T, Li, L, Pan, W and Liu, J (1996) 'Acute Health Effects of Major Air Pollutants in Taiwan', in Mendelsohn, R and Shaw, D (eds), *The Economics of Pollution Control in the Asia Pacific*, Edward Elgar, Cheltenham

Stern, N H (1977) 'Welfare weights and the elasticity of the marginal valuation of income' in M Artis, and A Nobay (eds) *Studies in Modern Economic Analysis: the Proceedings of the Association of University Teachers of Economics*, Blackwell, Oxford

Stymne, S and Jackson, T (2000) 'Intra-generational equity and sustainable welfare: a time series analysis for the UK and Sweden', *Ecological Economics, No 33*, pp219–236.

Symons, E, Proops, J and Gay, P (1994) 'Carbon Taxes, Consumer Demand and Carbon Dioxide Emissions: A Simulation Analysis for the UK', *Fiscal Studies*, Vol 15, No 2

Taylor, T, Markandya, A and Hunt, A (2001) 'Trade and Environment: Linkages in Multilateral Environmental Agreements', paper presented to workshop on Trade and Environment, FEEM, Milan, May.

Weitzman, M (1998) 'Gamma Discounting and Global Warming', discussion paper, Harvard University, Harvard, MA

Williamson, J G (1991) *Inequality, Poverty and History*, cited in Meier (1995)

Xu, Z (1994) 'Assessing Distributional Impacts of Forest Policies and Projects: An Integrated Approach', *Evaluation Review*, Vol 18, No 3, pp281–311
Zhou, P (1999) 'Climate change mitigation in Southern Africa', Botswana country study, Risø National Laboratory, UNEP Collaborating Centre on Energy and Environment, Roskilde, available online from www.uccee.org.

APPENDIX 6.1: ANALYSIS OF THE DISTRIBUTION OF CONTROL COST AND ANCILLARY BENEFITS

The distributional impacts of the different mitigation projects considered in Chapter 7 will depend on the method for raising the control cost and the distributional characteristics of any ancillary impacts. The case study presented in this chapter considered two possible scenarios where distribution may be important: where the project is self-funded through increased prices of GHG-related commodities and where the project is funded through general taxation. This appendix provides the background assumptions made in the earlier analysis.

Road-paving Scheme

The distributional implications of a self-funding road-paving scheme was assessed on the basis of raising a charge on petrol consumption as a proxy for road usage. The proportion of total expenditure on petroleum by different income groups was obtained from the Botswana HIES, and is shown in Table 6.15.

Efficient Lighting

The distribution of lighting was assessed in a survey by the Botswana Power Corporation (BPC, 1996). The results of this survey are presented in Table 6.14.

Low-cost housing thus has less access to lighting compared to the other two groups. In terms of cost, the efficient lighting scheme has been shown to carry a negative financial cost. This would suggest that the scheme could be self-funding. However, the distribution of the benefits of this project may be thought to be skewed towards the richer groups in Botswana. The cost of 40 pula for the more efficient light may be hard to raise for the poorer groups and the increased baseline education levels of the richer groups in Botswana would imply that implementation of this project would more likely fall on the richer groups. Also, the usage of lighting is greater in richer households, implying that the gains from reduced electricity consumption (amounting to P350,000) would fall more heavily on these groups. Thus, although this scheme would confer positive benefits to all groups with access, it is likely to cause greater inequality, so on a pure inequality cost measure the degree of inequality is likely to rise.

In terms of raising revenues to fund the project, an additional charge could be made on lightbulbs. The distribution of expenditure on lightbulbs was presented in the Botswana HIES, as shown in Table 6.16. It can be seen that the distribution of expenditure on lightbulbs rises with income, as the relative measure indicates (where the relative measure shows the proportion of expenditure divided by the proportion of households in each income category). Given

Table 6.14 *Distribution of lighting in Botswana*

Type of household	Number of lights	Usage (hours per day)
Low cost	6	4
Medium cost	12	4
High cost	23	4

Source: Based on BPC (1996) cited in Zhou (1999)

that the use of lightbulbs in terms of hours per day does not alter across income groups, as shown above, the expenditure on lightbulbs was also used as a proxy to assess the distribution of energy savings. This assumes that all parties take advantage of the new lightbulbs, which, as noted above, may not necessarily be the case.

Power Factor Correction

In the self-funding case, the distribution of the cost element of the power factor correction scheme was assumed to be the same as that for electricity consumption, hence it was assumed that a proportionate increase would be made to electricity prices to cover the cost. The distribution of electricity consumption was calculated from the Botswana HIES, and is shown in Table 6.17.

Pipeline Construction

For the construction of a new pipeline it was assumed that if the scheme were to be self-funded, a charge would be levied on petrol. Hence, the distribution was assumed to be the same as for the costs of road paving as shown above.

Central PV Plant

The construction costs of a central PV plant could be raised through increasing electricity prices. Thus, for the self-funded option it was assumed that the distribution of costs across income groups would follow that for power factor correction shown above.

Industrial Boilers

For industrial boilers, self-funded options affecting consumers were felt to be unlikely. As such, no measure was made for this option, although the case of central government funding through general taxation was considered.

Vehicle Inspection

In terms of road vehicle inspection, the distribution of vehicle ownership across income groups was reported in the Botswana HIES, 1996, and is shown in Table 6.18 below. As can be seen from the table, car ownership in general rises with income in Botswana, suggesting that measures to levy a charge on vehicle inspection would be largely progressive in nature.

Table 6.15 *Distribution of petroleum sales by income group*

	Consumption expenditure (pula/month)										
	<100	100–200	200–300	300–400	400–500	500–750	750–1000	1000–1500	1500–2000	2000–3000	>3000
Percentage of hh	20.65	16.76	11.71	9.39	5.36	10.23	7.32	6.4	3.99	3.59	4.59
Petrol and diesel oil expenditure	0.03	0.6	1.84	1.18	3.1	7.29	18.51	30.14	46.86	99.73	154.56
Proportion of total petrol consumption	0.0004	0.0059	0.0125	0.0065	0.0097	0.0434	0.0789	0.1123	0.0189	0.2085	0.4131

Table 6.16 *Consumption of lightbulbs by income group*

	Consumption expenditure (pula/month)										
	<100	100–200	200–300	300–400	400–500	500–750	750–1000	1000–1500	1500–2000	2000–3000	>3000
Proportion of total lightbulb expenditure	0.047	0.038	0.072	0.079	0.037	0.289	0.056	0.073	0.009	0.069	0.231

Table 6.17 *Distribution of electricity consumption in Botswana*

	Income group (pula/month)										
	<100	100–200	200–300	300–400	400–500	500–750	750–1000	1000–1500	1500–2000	2000–3000	>3000
Expenditure on electricity	0	0.09	1.12	0.84	1.51	1.51	5.45	16.32	14.87	36.17	109.17
Proportion consumed	0	0.001736	0.001617	0.009077	0.009314	0.0177768	0.04591	0.120199	0.068279	0.149433	0.576658

General Taxation

For all projects, funding based on raising general taxation was considered. Drawing on data from the Botswana HIES, the tax charged to different income groups was calculated based on the difference between disposable and gross income. The proportionate levels of revenue collected from each income group was used to calculate the impacts of project spending on different income groups. The data used for this analysis are presented in Table 6.19.

Ancillary Benefit Distribution

The distribution of ancillary benefits resulting from the mitigation projects examined was estimated where possible. The health benefits from reduced SO_2 emissions were assumed to be equally distributed, in the absence of data on the income levels of those living in the affected area. Hence a distributional weight of 1 was assumed, although, as suggested earlier, the distributional nature of air pollutants might be assumed to be regressive, but the WTP estimates would equally have to be adjusted for income levels, which might suggest that applying a weight of 1 is not unreasonable.

For employment impacts, these were assumed to have impacted the groups with income 30 per cent of the average owing to the distribution of income for unemployed labourers.[15] For mining health benefits, based on the analysis mentioned in Chapter 4, no weight was applied as miners have incomes similar to the average level in Botswana.

Table 6.18 *Vehicle ownership as percentage by income group in Botswana*

National	>100	100–200	200–300	300–400	400–500	500–750	750–1000	1000–2000	2000–3000	3000–5000	>5000	All
Number of Respondents	63292	36462	26898	27389	17856	32502	21571	36643	13832	8792	6372	291610
Motorcycle	0.20	0.20	0.40	0.30	1.10	0.90	0.00	0.90	1.90	2.10	6.60	0.70
Van	1.30	2.90	3.50	4.20	3.40	5.40	4.30	13.90	26.80	49.70	52.00	8.10
Car	1.50	1.10	0.50	3.10	3.50	3.10	3.50	13.50	21.70	51.40	71.10	7.40

Source: Botswana HIES, 1996

Table 6.19 *Distribution of tax revenues in Botswana: background data*

Proportion	Gross income								
	<100	100–200	200–300	300–500	500–1000	1000–1500	1500–2000	2000–3000	>3000
Tax	0.036141	0.045177	0.32139	0.032139	0.039006	0.038472	0.038	0.061088	0.0721
Population	0.217044	0.125037	0.09224	0.155156	0.18543	0.062829	0.062829	0.047433	0.052001
Revenues	0.009728	0.017751	0.015249	0.048938	0.106149	0.069772	0.085895	0.148126	0.498392

Source: Botswana HIES 1993–1994 and the author's own calculations

Chapter 7

Case Studies for Zimbabwe, Botswana, Mauritius and Thailand

Kirsten Halsnaes, Anil Markandya and Tim Taylor

INTRODUCTION

So far, the discussion in this book has focused on the methods of assessment of climate change mitigation projects and on the need for them to include the impacts of these projects on the environment and society more broadly. We noted in Chapter 5 that a number of decision-making rules are available, the choice between them depending on who the policy-maker represents. Developing country public sector representatives would find a benefit-cost approach, combined with the selected use of sustainability indicators, most helpful. The private sector, on the other hand, would look more narrowly at the cost-effectiveness of the projects, while donor agencies would be interested in a wider cost-effectiveness, as well as an estimate of the net benefits and a number of the sustainability indicators discussed in Chapters 2–4.

This chapter takes the cost-effectiveness approach and demonstrates how it would work in practice, by analysing in some detail the costs of implementing greenhouse gas (GHG) emission-reduction projects in Zimbabwe, Botswana, Mauritius and Thailand. The cost-effectiveness of the projects is seen from a financial as well as a social cost perspective. Financial costs represent the costs facing a project implementing agent, while the social costs include a broader range of impacts related to development, and the local environmental and social issues. The assessment of financial and social project costs leads up to a discussion about potential trade-offs and synergies between national development objectives and global cost-effectiveness of GHG emission-reduction policies. The approach is a further development of a methodological framework which was established as part of the UNEP/GEF project Economics of GHG Limitations (Halsnaes, Callaway and Meyer, 1998; Markandya, 1998). In making the social cost assessment the chapter also introduces the use of methods of valuation for environmental costs. Hitherto in this book we have assumed that such valuations are possible. There are, however, a number of problems with actually eliciting the numbers and, although we do not go into these in great detail, it is important to be aware of what the broad issues are and where the major controversies lie.

The case studies assessed are based on comprehensive national mitigation costing studies that have been conducted by national teams in collaboration with UNEP and ADB/UNDP for Zimbabwe, Botswana, Mauritius and Thailand. These studies focused on the assessment of the direct costs of implementing specific projects or sectoral strategies, including capital costs, fuel costs, land costs, maintenance costs, and implementation costs. The implementation of GHG emission-reduction projects, however, in addition to these direct costs will have implications on a broader range of national development policies, including employment generation, income distribution and local environmental policies.

The case studies have supplemented the direct cost assessment with information about side-impacts on local air pollution, employment generation and health impacts in the coal-mining sector. Specific national data have been used to estimate local employment impacts, while international data have been used to represent damages of local air pollution and health impacts in the coal-mining sector. In this way, national side-impacts have not been estimated accurately, and the social cost estimates established in this study, therefore, should be considered only as a first approximation of the costs. However, the established information is considered to provide a number of useful insights about potential trade-offs and synergies between direct GHG emission-reduction costs and the broader national impacts.

The studies show that the cost-effectiveness ranking of the GHG emission projects considered varies considerably between a ranking based on financial costs versus one based on social costs. Social costs, in almost all project cases, are assessed to be lower than financial costs, which indicates that the projects have a number of side-benefits seen in a national development perspective. A selection of GHG emission-reduction projects based on financial costs, therefore, will be different from a project selection based on social costs, and the magnitude of the project costs also will vary significantly with the cost-assessment perspective.

The chapter starts with a short introduction of the methodological framework followed by definitions of the financial and social cost concepts applied to the project assessment. The framework for assessing employment impacts, local air pollution impacts and health impacts in the coal-mining sector as part of the social costs are summarized, and the specific values for health and pollution damage estimates used in the country studies are reported. Financial and social costs are assessed for GHG emission-reduction case projects in each of the countries, and the impacts of including employment generation, local air pollution and coal-mining impacts in the costs are discussed for the project cases. Finally, a number of general cross-cutting discussions about the differences between a project selection based on financial and social costs are included.

METHODOLOGICAL FRAMEWORK

Cost Concepts

Actions taken to reduce GHG emissions will generally divert resources from other alternative uses. The cost assessment should ideally consider all changes in resources demanded and supplied by a given GHG emission-reduction project or strategy in relation to a specific non-policy case. The assessment should include, as far as possible, all resource components and implementation costs, and all costs as well as benefits of the project.

The underlying objective behind any cost assessment is to measure the change in human welfare generated as a result of a reallocation or use of resources. This implies the existence of a function in which welfare or 'utility' depends on various factors such as the amounts of goods and services that the individual can access, different aspects of the individual's physical and spiritual environment, and his or her rights and liberties. Constructing a 'utility-function', representing social welfare which is an aggregate measure of all such impacts for all individuals, involves a number of complexities and controversial equity issues that have been studied intensively by economists. To a considerable extent these difficulties can be finessed by taking the sum of the individual willingness to pay/ willingness to accept (WTP/WTA) as a measure of the social welfare. The authors have selected in the present assessment a number of indicators of welfare which have been valued as the sum of the individuals WTPs/WTAs. These indicators have been measured in monetary values, taking the view that the methods for 'converting' some of these other dimensions into monetary terms are useful and should be pursued.

The costs of project implementation basically can be assessed from two different perspectives – namely, from a social and a private perspective.

The aim of a social cost assessment is to include all costs and benefits of a project implementation, including impacts that occur to private agents that are directly involved in the project implementation, as well as impacts on others of his or her actions. These latter costs can be referred to as externality costs.

Financial costs are in the context of the current assessment defined to reflect the costs that a private agent faces in implementing a given project. The financial costs reflect the actual financial transfers in the form of capital, operation costs, planning and training, and other implementation expenditures that are needed to get a GHG emission-reduction project into operation.[1]

The current study is a further development of a methodological framework for the assessment of indirect costs and benefits developed by Markandya (1998) as part of the United Nations Environment Programme (UNEP) project Economics of GHG Limitations. The specific indirect costs and benefits that have been included in the social cost assessment are:

1   Employment impacts.
2   Health impacts.
3   Associated environmental changes.

A number of specific methodological issues and assumptions applied to the assessment of employment impacts, health impacts and associated environmental changes are outlined in the following.

Evaluating Employment Impacts

The creation of jobs in relation to project implementation is a benefit to society to the extent that the persons employed do not presently have a job. The social benefit of the employment, then, is equal to the social costs of the unemployment avoided. These benefits will depend primarily on the period that a person is employed, what state support is offered during any period of unemployment, and what opportunities there are for informal activities that generate income in cash or kind. In addition, unemployment is known to create health problems which have to be considered as part of the social cost.

Before setting out the framework for such an evaluation, it is important to set out the theoretical reasons for arguing that unemployment reduction has a social value. In neoclassical economic analysis, no social cost is normally associated with unemployment. The presumption is that the economy is effectively fully employed and that any measured unemployment is the result of matching the changing demand for labour to a changing supply. In a well-functioning and stable market, individuals can anticipate periods when they will be out of work, as they leave one job and move to another. Consequently, the terms of labour employment contracts, as well as the terms of unemployment insurance, will reflect the presence of such periods, and there will be no cost to society from the existence of a pool of such unemployed workers.

However, these conditions are far from the reality in most of the developing countries in which the GHG projects will be undertaken. Many of those presently unemployed have bleak prospects of finding stable employment. In general, unemployment is a primary worry among those who are presently employed, and the political pressure not to take measures that will further increase this level is very high.

On this basis, the current study has treated the welfare gain of those made employed as a social gain. Traditionally, this welfare gain has been defined as:

(a) the gain of net income as a result of a new job, after allowing for any unemployment benefit, informal employment, work-related expenses, and so on, minus
(b) the value of the additional time that the person has at his or her disposal as a result of being unemployed and that is lost as a result of being employed, plus
(c) the value of any health-related consequences of being unemployed that are no longer incurred.

To calculate the social benefits (the unemployment avoided as a result of the project), the welfare cost is multiplied by (a) minus (b) plus (c).

The social benefit assessment of increased employment includes the net salary increase (including income taxes) that a newly employed person can gain. The studies for Zimbabwe and Botswana assumed that the employment generation only included unskilled labour that was temporarily employed in relation to the implementation of a given project, and it was assumed here that no income tax was paid by those employed. Furthermore, it was assumed that no unemployment benefit system was in place, implying that the persons were only living on the basis of informal income-generating activities. The study for Mauritius assessed the employment benefits on the basis of income tax statistics and average unemployment benefits for the sectors affected by the project implementation (Markandya and Boyd, 1999).

In gaining employment an individual loses leisure time, which has some value depending on the elasticity of the labour supply. Few specific studies exist for developing countries and from these it has been assumed, conservatively, that the loss of leisure time has a value of 15 per cent of the gross wage rate.

The health impact associated with increased employment is particularly related to increased life expectancy among men in the employable age. Studies available for industrialized countries show that there is an excess mortality of around 75 per cent, with a range from 45 to 110 per cent for the unemployed (Markandya, 1998, p26). Despite the fact that no empirical estimates are available yet for developing countries, a value of 75 per cent excess mortality of men in the employable age has been applied to the assessment of the social benefits of the employment generation in the studies.

Valuation of Health Impacts

The valuation of health impacts has included estimates of changes in the risk of mortality as well as the risk of injury. The value of changes in the risk of mortality has been assessed on the basis of the value of statistical life (VOSL) methodology. The method converts the WTP/WTA for a certain risk of death into a VOSL in the following way. Suppose a group of individuals is willing to pay US$2 each year to reduce the annual risk of death from some external factor (such as air pollution) by one in a million. This can be interpreted to mean that if the risk is reduced, a population of 1 million will face one less loss of life and at the same time a payment of US$2 million will have been made. Hence US$2 million is called the VOSL. Of course, there are many issues that arise in making this calculation. How will it change with the risk of death, with the population under risk and with the nature of the risk? For an in-depth discussion of these issues, see Markandya (1996).[2]

For the European Union (EU), Markandya estimated a central VOSL at US$3.9 million in 1995 prices. Unfortunately, studies do not exist for most developing countries, although work is ongoing in this area. Until we have information from such primary studies, a common procedure is to carry out a 'benefit transfer' which entails making an adjustment to the value from one country or application to another. For example, the VOSL for the EU can be adjusted to arrive at one for, say, India. The most important factor in making this adjustment is to allow for the fact that the WTP for risk reductions will be much lower, given the lower individual incomes. If incomes in India are, say, 20 per cent of those in the EU, how much lower should the VOSL be? It depends on a factor referred to as the income elasticity of WTP for a reduced risk of death. If that elasticity is one, the VOSL should also be 20 per cent lower. There are studies on the income elasticity for some environmental goods but none for mortality (see Markandya, 2001). One might argue that the elasticity is greater than one if environmental goods and risk reduction are seen as something of a luxury (expenditure rises with income but by a greater percentage than income). In practice, however, the evidence is not supportive of an elasticity of more than one, and a 'default' value of one is not inconsistent with some of the evidence (for a discussion of this issue, see Markandya, 2001). Hence, in what follows a value of one has been adopted. This implies that the VOSL has been estimated for non-EU countries adjusting the EU VOSL with the ratio of real per capita income in the country concerned to the gross domestic product (GDP) in the US, based on the purchasing power parity (PPP) GDPs in US$ 1994 reported by the World Development Indicators (WDI, 1998). The VOSL values for the case study countries are given in Table 7.1.

Table 7.1 *VOSL values applied to national health impact estimates*

	VOSL in PPP GNP 1995 US$1000
Zimbabwe	315
Botswana	1085
Mauritius	1238
Thailand	881

The VOSL values are very sensitive to annual fluctuations in PPP exchange rates. It can be seen, for example, in Figure 7.1 that the PPP exchange rate per capita GDP for Botswana varies considerably between 1993 and 1996, which will be reflected in VOSL values that are based on data for one year (WDI, 1998). This is a problem particularly in cross-country comparisons of VOSL-based estimates, where it can be appropriate to use a forecast of VOSL values showing the likely development trend over a longer period.

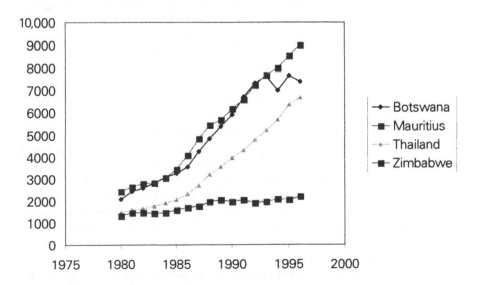

Figure 7.1 *GNP per capita, PPP in US$*

Occupational health impacts for the coal-mining sector have been included in the studies for Zimbabwe and Botswana. The occupational health impact data are based on statistics for coal-mining activities in South Africa, which must be expected to have a lower level of risk than the mining sector in Zimbabwe and Botswana. See Appendix 7.1 for an extensive description of methodological framework and data applied to these studies on occupational health impacts.

The following health impact categories have been included for the coal-mining sector:

- Risk of mortality from accidents.
- Risk of serious injury.
- Risk of minor injury.
- Risk of pneumoconiosis.

The valuation of the risk of serious and minor injury as well as pneumoconiosis include the expenditures of medical care and human costs from suffering plus loss of income emerging from sickness. These costs have been assessed on the basis of data from the UK that have been included in the ExternE project (ExternE, 1999), adjusted to reflect the differences in PPP GDP level between the countries (WDI, 1998). Again, this can only be a rough approximation. In practice, actual data on these costs needs to be collected at the national level.

Environmental Impacts

GHG emission-reduction projects in many cases will have impacts on the local environment in relation to air quality, water quality and waste generation. The current project study has only included an assessment of the impacts on local air quality in the form of health impacts related to SO_2, NO_x and particulate emissions.[3] The following national case study subsections will briefly discuss the impacts of excluding other relevant environmental impacts.

The health damages of SO_2, NO_x and particulate emissions have been assessed in various studies for the US, Europe and other industrialized countries. Studies include ExternE 1995 and 1999 for Europe, Rowe et al (1995) for New York, Thayer et al (1994) for California, and a number of studies for developing countries (Shah, Nagpal and Brandon , 1997; Dessus and O'Connor, 1999; Pearce, 1996). The actual damages of these emissions are site and time dependent, and accurate values of these can only be established on the basis of project-specific pollution impacts and damages studies that have not been available for the countries considered. The current assessment has therefore used a range of damage estimates[4] for the US and European studies and adjusted them on the basis of differences in real per capita GDP. It also implies that the same air pollution damage values have been applied to the emissions originating from the GHG project and those substituted in the baseline case. Values calculated in this way should be considered as highly uncertain, but can give an indication of the range of damages avoided.

The application of this adjustment methodology has been considered to impose specific uncertainties in the case of Mauritius, which is a small island state with a high population density. The total population of Mauritius that will be exposed to SO_2, NO_x and particulate emissions must be considered to be relatively large compared with the population exposed in the countries included in the ExternE studies. The damage values for local air pollution for Mauritius have been scaled therefore with a population density factor[5] (Markandya and Boyd, 1999).

Zimbabwe, Botswana and Thailand have a lower population density than the UK and Germany, on which the ExternE data is based. However, the actual air pollution exposure to local people has been assessed to be relatively high in Zimbabwe, Botswana and Thailand. This is the case because the emission sources with high SO_2, NO_x and particulate emissions are located to a large extent in areas with high population density (ALGAS, 1999a; Department of Mines Botswana, 2000). The ExternE data have therefore been considered as a useful first approximation of local air pollution damages. The adjusted air pollution values that have been applied to measure the benefits of improved air quality in Zimbabwe, Botswana, Mauritius and Thailand are shown in Table 7.2.

Table 7.2 *Values of air pollution damages for SO_2, NO_x and particulates used in the case studies (US\$ 1996 per tonne)*

	SO_2 low estimate	SO_2 high estimate	NO_x low estimate	NO_x high estimate	PM low estimate	PM high estimate
Zimbabwe	1070	1407	554	1424	1769	4684
Botswana	3688	4851	879	2263	2809	7439
Mauritius electricity generation	–	7053	–	2357	–	14,677
transport		7383	–	8970	–	15,386
Thailand	2989	3931	1547	3980	4944	13,087

Note: The values are based on ratios of national 1994 PPP GNP to the US 1994 PPP GNP

The results subsequently reported in the national study sections have used the high pollution damage values.

DEFINITION OF THE COST-EFFECTIVENESS CONCEPTS APPLIED TO THE CASE STUDIES

The cost assessment considers the financial and social costs of GHG emission-reduction projects based on case examples for Zimbabwe, Botswana, Mauritius and Thailand. The costs are assessed on the basis of a comparison of the policy case with a baseline case. The baseline case by definition reflects the national policy priorities, given that GHG emission reduction is not an objective. The baseline case is defined in relation to a specific project that is assumed to be substituted by the GHG emission reduction over the lifetime of the project.[6] The baseline project includes assumptions about all data that is assessed for the GHG emission-reduction project, including data on energy consumption, costs, GHG emissions, employment, and local air pollution. This implies that the costs are measured as a comparative assessment of the specified parameters in the policy and baseline cases.

An overview of the specific baseline projects assumed in each of the case studies is given subsequently for each of the countries in Boxes 7.2–7.5.

The financial cost assessment includes two concepts – namely, gross and net financial costs. Gross financial costs are the costs of running the GHG emission-reduction project, excluding the costs of the substituted baseline activity. Net financial costs are the financial costs of a given project measured as an incremental cost in relation to running a baseline case. The difference between the net and gross financial cost concept can be illustrated in the case of an efficient industrial boiler as a GHG emission-reduction project. The gross financial costs of the project will be the capital cost of the motor, the cost of operation and maintenance, including power, and implementation costs. The net financial costs for this project will be the gross financial costs minus the savings in operation and maintenance cost, including the value of electricity savings assessed in relation to the baseline case.

The social cost assessment includes two concepts – namely:

- Social cost alternative 1 where national co-benefits on local air pollution, time savings and employment have been included.
- Social cost alternative 2 that, in addition to the impacts included in alternative 1, also includes health impacts in the domestic coal-mining sector.

The GHG emission reductions achieved through the projects are both in the gross and net financial cost cases and in the social costs cases assessed as the difference between the GHG emissions in the policy case and the emissions in the baseline case.

The cost-effectiveness of the GHG emission-reduction projects in meeting the GHG emission-reduction objective is calculated for financial and social costs. The financial and social costs in most cases include costs as well as benefits, and the benefits are in the cost-effectiveness analysis 'counted' as a negative cost.

The costs of the reduction projects, $Costs_{Reduction}$, are calculated as a summarization of all costs and benefit components of a project, j at the time t;

$$Costs_{Reduction_t} = \sum_{j=1}^{n} Costs_{Reduction_{jt}}$$

and the costs of the baseline projects, $Costs_{Baseline}$, are calculated similarly as:

$$Costs_{Baseline_t} = \sum_{j=1}^{n} Costs_{Baseline_{jt}}$$

The costs are defined at a financial and social cost basis, and these costs are labelled respectively *FiCosts* and *SoCosts*.

The *GHG reduction* is defined as the difference between the GHG emissions in the CDM project case and the baseline case, at the time t, as follows:

$$GHG_{Reduction_t} = GHG_{Baseline_t} - GHG_{Project_t}$$

The cost-effectiveness of the projects is defined as the ratio of the net present values of the costs and the GHG emission reduction. See the section on GHG emission reductions (pp212–213) for a more detailed discussion about the discounting of emission reduction.

The cost-effectiveness of the projects is calculated in relation to four reduction-cost concepts gross financial costs, net financial costs, social costs alternative 1 and social costs alternative 2.

Gross financial costs, GFI, are defined as:

$$GFI = \frac{\sum_{t=0}^{T} \dfrac{FiCosts_{Reduction_t}}{(1+i)^t}}{\sum_{t=0}^{T} \dfrac{GHG_{Baseline_t} - GHG_{Project_t}}{(1+r)^t}}$$

where $FiCosts_{Reduction}$ are the financial costs of the reduction projects which are summarized over the j cost and benefit components of the reduction project and the baseline project respectively. It should be noticed that the gross financial costs as defined here include the full costs of project implementation without subtraction of the eventual cost savings from the substituted baseline activity. The GHG emission reductions of the project, however, are calculated as the difference between the GHG emissions in the reduction project case and in the baseline case. The rationale for this gross financial cost concept is that these costs reflect the actual financial costs facing private agents that are implementing the project.

Net financial costs, *NFI*, are defined as:

$$NFI = \frac{\sum_{t=0}^{T} \dfrac{FiCosts_{Reduction_t} - FiCosts_{Baseline_t}}{(1+i)^t}}{\sum_{t=0}^{T} \dfrac{GHG_{Baseline_t} - GHG_{Project_t}}{(1+r)^t}}$$

where $FiCosts_{Reduction}$ and $FiCosts_{Baseline}$ are the financial costs of the reduction project and the baseline project summarized over the j costs and benefit components of the reduction project and the baseline project respectively. Net financial costs differently from gross financial costs, assess the financial costs as the difference between the reduction project case and the baseline case, implying that eventual cost savings incurred by substituting the baseline case are subtracted. Such baseline case cost savings can be related, for example, to fuel consumption, maintenance and capital costs. Costs savings measured in relation to the substituted baseline case activity can occur to the project implementing agent, other private agents or the public authority. GHG emission reduction is calculated as the difference between the emissions in the project and baseline cases.

Social costs, *SOC*, are defined as:

$$SOC = \frac{\sum_{t=0}^{T} \dfrac{SoCosts_{Reduction_t} - SoCosts_{Baseline_t}}{(1+i)^t}}{\sum_{t=0}^{T} \dfrac{GHG_{Baseline_t} - GHG_{Project_t}}{(1+r)^t}}$$

where $SoCosts_{Reduction}$ and $SoCosts_{Baseline}$ are the social costs of the reduction project and the baseline project summarized over the j costs and benefit components of the reduction project and the baseline project respectively. *SOC* is calculated for social cost alternatives 1 and 2. The social costs in all cases are assessed as the difference between the costs of the project and baseline cases, and the same principle applies to the GHG emission reductions.

GHG Emission Reductions

The GHG emission reductions occur at different points in time, and time-specific values of the avoided damages that these GHG emission reductions generate therefore can have important implications on the cost-effectiveness of the policies.

There is large uncertainty about climate change damages and it is difficult therefore to assign time-specific values to the damages avoided by reduced GHG emissions at different points in time.[7] It has therefore for simplicity in the current study been assumed that the avoided climate change damage per unit of GHG emission is constant over the timeframe considered.[8] Following this approach, GHG emission reduction can be considered as a proxy for the avoided climate change damages. These proxy 'damages' (measured by the GHG emission reductions) can then be considered as an indicator of the relative performance of the projects in meeting the GHG emission-reduction objective. The proxy damages are then transferred to comparable time-independent values by discounting to net present values in the same way as for the project costs.

Discount Rates

The choice of discount rates applied to the cost and GHG emission-reduction calculations depends on the perspective of the analysis. The discount rate used in the cost calculations can vary with the financial and social perspectives.

A financial cost analysis by definition reflects the costs of implementing a project seen from the perspective of a private agent, and the costs therefore should reflect the actual cost of financing the project. This discount rate can be the market interest rate if the project is financed via the local market, or it can be interest rates to be paid on international loans if that is the financial source. It must be considered to be most likely that GHG emission-reduction projects at present will be initiated primarily in developing countries if they are internationally financed and the current assessment has therefore based the financial cost assessment on the discount rate used in international financing studies. The rates used by international banks in appraising investment projects in developing countries are around 10–12 per cent.

For social cost calculations, it is relevant to base the assessment on a discount rate that reflects the opportunity cost of capital. In developed countries, rates of around 4–6 per cent would probably be justified. Rates of this level are used, in fact, for the appraisal of public sector projects in the EU (Watts, 1999). In developing countries the rate could be as high as 10–12 per cent. The social cost assessment has therefore used a 10 per cent social discount rate as a central value.

The GHG emission reductions, as discussed previously, have been transformed to a proxy of the NPVs of the avoided damages. The climate change damages can be understood as a social benefit[9] that occurs over a very long time horizon. The discount rate that should be applied to this specific benefit has been discussed extensively in the literature (IPCC, 1996, Chapter 5; Portney and Weyant, 1999). Some authors argue that the appropriate discount rate for climate change damages in any time frame is the opportunity cost of capital; others argue that the discount rate should be much lower for very long time horizons to reflect intergenerational equity concerns. The current project assessment is considering a relatively short time horizon (up to 30 years) and it has therefore been considered to be reasonable to use a discount rate that reflects the opportunity costs of capital. This discount rate has used similarly to the cost assessment 10 per cent as a central value. Appendix 7.3 includes the results of a sensitivity analysis where a discount rate of 3 per cent has been used to calculate a proxy NPV of the GHG emission reductions.

CASE STUDY OVERVIEW

The case studies are based on detailed national mitigation costing studies that have been carried out as part of large international country study programmes conducted by the UNEP and the Asian Development Bank in collaboration with the United Nations Development Programme (UNDP). Appendix 7.2 provides details of the mitigation options evaluated for each country.

The study for Zimbabwe has been part of the UNEP National Abatement Costing Studies that was carried out in the period from 1992 to 1994 and included studies for Brazil, Venezuela, Senegal, Zimbabwe, Egypt, India and Thailand (UNEP, 1994; Southern Centre, 1993, 1995). These studies focused on the assessment of the abatement costs of energy sector projects.

The studies for Botswana and Mauritius were part of the UNEP/GEF project Economics of Greenhouse Gas Limitations that, in addition to the studies of these two countries included studies for Ecuador, Argentina, Senegal, Zambia, Tanzania, Vietnam, Indonesia, Estonia and Hungary. The studies were based on a common methodological framework providing definitions of key concepts, analytical structure and technical assumptions for the assessment of projects and sector strategies for the energy sector, forestry, agriculture, waste management and the transportation sector. The studies in general focused on the assessment of the direct costs of GHG emission reduction, but also included the development of a methodological framework for assessing indirect costs and benefits of the policies.

This framework was applied to studies for Mauritius and Hungary (Markandya and Boyd, 1999; Zilahy et al,1999). These indirect cost and benefit studies have been supplemented in the context of the current paper with a study for Botswana (UNEP, 1999) and a study for Thailand. Thailand was part of an ADB/UNDP programme that, in addition to this study, included Bangladesh, The People's Republic of China, India, Indonesia, The Republic of Korea, Mongolia, The Union of Myanmar, Pakistan, the Philippines and Vietnam.

The methodological approach of the UNEP and ALGAS studies has been to use a number of different sector models and to use the models to provide a number of key results about the costs of meeting specific emission-reduction targets at a given point in time (Halsnaes, Callaway and Meyer, 1998; ALGAS, 1999a). The energy sector analysis has been conducted typically with integrated energy system models, including various power sector and end-use energy options. This implies that the energy sector projects included in the current study are assessed in relation to a baseline case that is established as part of a more comprehensive energy sector strategy.

The current assessment of financial and social costs has included a subset of the options that originally were included in the studies for Zimbabwe, Botswana, Mauritius and Thailand (see Appendix 7.2 for details). A priority has been given to options with relative high GHG emission-reduction potential and low direct costs. Large power supply options have not been included in the financial and social cost assessment because it has been considered that such projects could have very significant indirect impacts that should be assessed in more detailed site specific studies.

An overview of the case examples assessed for the countries are given in Box 7.1, which later is supplemented with more details on the individual projects in the sections on the specific case studies.

Box 7.1 Overview of case examples

Zimbabwe	*Botswana*
• Biogas for rural kitchens	• Road pavement
• Efficient lighting	• Efficient lighting
• Efficient furnaces in manufacture	• Power factor correction
• Efficient tobacco barn	• Pipeline for fuel oil
• Industrial boilers	• Central PV plant
• Sewage biogas for electricity	• Industrial boilers
	• Vehicle inspection
Mauritius	*Thailand*
• Wind turbines	• Industrial motors
• LPG buses	• Biomass cogeneration
• Solar water	• Fuel efficiency in transport
• PV street lights	• Efficient stoves to reduce
• Bagasse power	woodfuel consumption

COUNTRY CASE STUDY RESULTS

The current section reports the main assumptions used in assessing the financial and social costs of GHG emission-reduction projects in Zimbabwe, Botswana, Mauritius and Thailand, and reports the results of the cost assessment for each of the countries, and the cost-effectiveness implications, considering the financial versus the social costs, are discussed.

Zimbabwe Case Study

The GHG emission-reduction case projects assessed for Zimbabwe have been selected on the basis of mitigation costing studies that cover the energy sector and a number of options for agriculture and forestry conducted as a collaboration between the Southern Centre for Energy and Environment and the UNEP Collaborating Centre on Energy and Environment (Southern Centre, 1993, 1995). The Zimbabwean studies have assessed over 20 mitigation options covering a 30-year timeframe, including options with a CO_2 reduction potential and options that are more targeted to reduce CH_4 emissions and N_2O emissions (see Box 7.2 and Appendix 7.2 for an overview of the options).

The GHG emission-reduction projects included in this study are a number of energy efficiency options in the household sector and industry, and biogas use in households and for power production. Two electricity options are included – namely, efficient household lighting and an efficient coal-based furnace for nickel and copper production that substitutes an existing electric furnace. Other industrial options included are efficient coal-fired boilers and an efficient tobacco barn. All these options imply decreased coal consumption and thereby have a number of local emission-saving impacts in addition to indirect impacts in the form of decreased demand for coal supplied from the Zimbabwean coal-mines. It should be noticed that Zimbabwean coal has a very high sulphur content (2.5 per cent) which makes the potential benefits of SO_2 emission reduction particularly high.

The case projects also include a sewage biogas project that reduces CH_4 emissions from the sewage plant and substitute coal-based electricity.[10] This project is assumed to substitute coal-based electricity production and therefore to have the same impacts on local air pollution and demand for locally mined coal as the industrial options and the electricity-saving options. Finally, the case studies include a biogas plant for rural households, which has a low financial cost and a large benefit in the form of time savings for local households due to the reduced need for woodfuel collection.

It can be seen from Table 7.3 that there is a significant difference between gross financial costs, net financial costs and social costs of the projects considered. The gross financial costs of the projects range from around US$4 per tonne of CO_2 to US$68 per tonne of CO_2, with the efficient tobacco barn as the cheapest project and efficient furnaces in industry as the most expensive. Net financial costs are in general lower than gross financial costs and for all projects are negative, with the biogas for rural kitchens project[11] as the only exception. The negative net financial costs are achieved because the GHG emission-reduction projects considered are assumed to substitute very inefficient baseline alternatives that are characterized by very high fuel consumption. Such baseline assumptions

Box 7.2 Characterization of case examples for Zimbabwe[1]

Case example	Project	Baseline case	Assumptions made in adjusting financial costs to get social costs
Biogas for rural kitchens	Biogas plant for anaerobic digestion of municipal and industrial waste. The plant produces CH_4 for cooking	Woodfuel for cooking. The woodfuel consumption implies deforestation	The social costs include the benefit of reduced SO_2, NO_x and particulate emissions
		CH_4 emissions from the waste	Time savings due to decreased woodfuel collection are valued
Efficient lighting	Introduction of 7-watt compact fluorescent lamps	40-watt incandescent lamps	The social costs include the benefit of reduced SO_2, NO_x and particulate emissions
		Coal-fired electricity	Health impacts implied by decreased coal production in the mines
Efficient furnace for Ni and Cu production plant	Installation of new efficient coal-fired furnace (coal-fired plasma arc technology based on an Australian design Siro-smelt plant)	Old electric furnaces based on coal-based power production	The social costs include the benefit of reduced SO_2, NO_x and particulate emissions
			Health impacts implied by decreased coal production in the mines
Efficient tobacco barn	New efficient barn fired with coal that reduces coal consumption per unit tobacco production by 68%	Old barn fired with coal	The social costs include a benefit of reduced SO_2, NO_x and particulate emissions
			Health impacts implied by decreased coal production in the mines
Industrial boilers	Improved coal-fired boiler with 79% efficiency	Existing coal-fired boiler with 50% efficiency	The social costs include a benefit of reduced SO_2, NO_x and particulate emissions
			Health impacts implied by decreased coal production in the mines
Sewage biogas for electricity	Use of the gas in a turbine or converted diesel engine to produce electricity	Coal-fired electricity	Social costs include a comparative assessment of the SO_2, NO_x and particulate emission of the biogas plant compared with the power plant
		CH_4 emissions from the biogas plant	
			Health impacts implied by decreased coal production in the mines

1 Detailed information about the options are included in national study reports for Zimbabwe by the Southern Centre for Energy and Environment, Zimbabwe; Department of Energy, Ministry of Transport and Energy, Zimbabwe; and System Analysis Department, Risø National Laboratory, Denmark, 1993 and 1995.

reflect the fact that the capital market in Zimbabwe is constrained, and the manufacturing sector to a large extent uses very old production facilities. A number of specific issues related to energy use in the manufacturing sector in Zimbabwe is discussed by Halsnaes (1997).

The estimates of the financial and social costs for the GHG emission-reduction projects in Zimbabwe are shown in Table 7.3.

Table 7.3 *Financial and social costs of GHG emission-reduction projects in Zimbabwe (costs are in 1990 US$[1] per tonne CO$_2$ reduction, 10 per cent discount rate)*

	Gross financial costs	Net financial costs	Social costs alternative 1	Local air pollution impacts[2]	Social costs, alternative 2	Coal-mining impacts
Efficient tobacco barn	3.8	−5.9	−63.3	−26.7	−107.5	−75.0
Biogas for rural kitchens	5.3	5.3	−24.6	−0.9	−24.6	0
Sewage biogas for electricity	12.4	−19.6	−119.4	−28.3	−119.4	−77.6
Efficient industrial boilers	16.1	−3.1	−31.4	−30.5	−109.0	69.9
Efficient lighting	44.3	−6.8	−41.1	−32.0	−105.9	−64.8
Efficient furnaces in manufacture	57.8	−8.1	−44.9	−35.7	−128.8	−94.9

1 Based on an exchange rate of 8.2 Zimbabwe dollars for one US dollar
2 A negative number means that the project is assessed to reduce local air pollution and thereby generates a benefit

Social costs are lower than financial costs for all the projects considered. The social costs in alternative 1 are negative for all projects and range from −US$119 per tonne of CO$_2$ for the sewage biogas plant to −US$25 per tonne of CO$_2$ for the biogas for rural kitchens project. These very low social costs reflect that very large side-impacts on local air pollution have been assessed for most of the options. The local air pollution impacts are assessed to be −US$35.7 for the efficient furnaces per tonne of CO$_2$ reduction and for the other options are also assessed to be as high as around −US$30 per tonne of CO$_2$. The high benefit values for most of the projects arrive because the projects substitute coal-based electricity or coal-fired boilers or furnaces that have very high local pollution impacts.[12] The biogas for rural kitchens project has here been assessed to have a very low side-impacts on local air pollution, but this result might change if more specific air pollution data for woodfuel use for cooking was included in the assessment. The social costs of the biogas project, however, are significantly lower than the financial costs because these options have been assessed to have significant time-savings benefits due to substituted woodfuel collection.

The social cost in alternative 2 is assessed to be lower than in alternative 1 because a number of the options are assumed to decrease the supply of coal from Zimbabwean mines and thereby reduce the negative health impacts on coal-mining workers. This health impact ranges from −US$95 per tonne of CO_2 for the efficient furnaces down to −US$65 per tonne of CO_2 for the lighting option. The reduced health impacts associated with reduced coal production can be somehow offset if the reduced coal production pushes coal-mining workers into unemployment. The biogas option is not related to coal consumption and therefore has a similar social cost in alternatives 1 and 2.

The cost-effectiveness ranking of the projects change from the financial cost cases to the social cost cases as shown in Table 7.4. The cost-effectiveness ranking of the projects based on gross financial costs primarily reflects the magnitude of the capital costs per unit of GHG emission reduction of the different projects, because they are all expected to have low fuel and maintenance costs. The cost-effectiveness ranking change is considerable when the perspective is changed to be net financial costs. In particular, projects that substitute power production move up in priority order over options that have smaller fuel-saving benefits.[13] The highest priority options now become projects that substitute coal-based electricity production, such as the sewage biogas project and the efficient lighting options. The next priorities according to net financial costs are furnaces, the tobacco barn and boilers.

The inclusion of local air pollution impacts such as in social cost alternative 1 makes the efficient tobacco barn particular more attractive in the cost-effectiveness ranking, compared with the ranking based on net financial costs, because this project implies a significant reduction of local air pollution. The most attractive project, in the social cost alternative 1 is still the sewage biogas project, and the two projects with the lowest ranking are industrial boilers and biogas for rural kitchens.

Table 7.4 *Ranking of GHG emission-reduction projects according to cost-effectiveness for Zimbabwe (the project with the lowest costs per tonne of CO_2 reduction has been given a ranking value of one)*

	Gross financial costs	Net financial costs	Social costs alternative 1	Social costs alternative 2
Efficient tobacco barn	1	4	2	4
Biogas for rural kitchens	2	6	6	6
Sewage biogas for electricity	3	1	1	2
Efficient industrial boilers	4	5	5	3
Efficient lighting	5	2	4	5
Efficient furnaces in manufacture	6	3	3	1

This ranking changes again when the coal-mining sector is included, as in the social cost, alternative 2. All projects that imply large decreases in coal consumption now become the most cost-effective, and they are furnaces in manufacture, the sewage biogas plant and industrial boilers.

Figure 7.2 shows marginal emission-reduction cost curves based on gross financial costs and the social cost, alternative 1. The projects here are depicted in ranking order according to cost-effectiveness measured as US$ per tonnes of CO_2 reduction.

The general conclusion that can be drawn from the gross financial cost curve is that the projects considered in total provide a basis for an NPV of about 70 million tonnes of CO_2 reduction for a cost below US$58 per tonne of CO_2. All the projects will potentially generate social benefits in Zimbabwe as a consequence of fuel savings and reduced local air pollution. The project with the largest potential social benefits are the efficient tobacco barn which actually are the most cost-effective project both in the gross financial cost perspective and in the social cost perspective. Attention should also be paid to the efficient furnace project, which offers a CO_2 equivalent emission-reduction potential as large as 45 million tonnes CO_2 for very low social costs. This project, however, is the lowest priority according to the gross financial cost ranking.

Botswana Case Studies

The case studies for Botswana have been selected on the basis of a comprehensive national climate change mitigation costing study covering the energy and transport sectors. Some of these have already been introduced in Chapter 5 to demonstrate the use of different decision-making tools. The list analysed here is, however, much wider.

The study was conducted as a collaboration between the Botswana Ministry of Minerals, Energy and Water Affairs, the local consultants EECG, and the UNEP Collaborating Centre on Energy and Environment (UNEP, 1999). The study included an assessment of 20 mitigation options covering a 30-year time horizon. The study assessed a total potential of annually 3 million tonnes of CO_2 for the projects.

The GHG emission-reduction projects that have been included in the current assessment include energy sector and transportation options. The transportation options are a road-pavement and a vehicle-inspection project. The projects also include electricity savings from efficient lighting, a central PV plant and power factor correction that increases the efficiency of the power transmission system. Industrial boiler efficiency improvements are as in the case of Zimbabwe assessed, and finally the study includes a pipeline project for transport of petroleum products from Pretoria to substitute road and rail transport of the fuel. The transportation projects and the pipeline projects all have a very large GHG emission-reduction potential and in the country study for Botswana assessed to contribute about 25 per cent of the total reduction potential in 2030. More details about the projects in the GHG emission-reduction case projects are given in Box 7.3.

The power production in Botswana is based on coal, and all options related to power production, therefore, will have indirect impacts on local air pollution

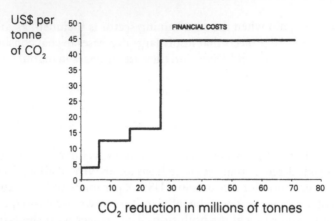

CO$_2$ reduction in millions of tonnes

Project	Gross financial costs US$ per tonne CO$_2$	CO$_2$ reduction million tonnes
Efficient tobacco barn	3.8	6.03
Biogas for rural kitchens	5.3	0.05
Sewage biogas	12.4	10.41
Industrial boilers	16.1	10.17
Efficient lighting	44.3	0.19
Efficient furnace	57.8	44.50

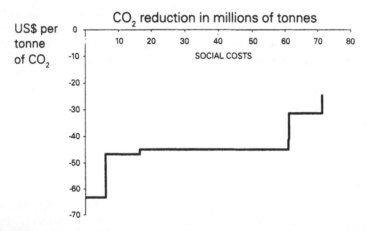

Project	Social costs US$ per tonne CO$_2$	CO$_2$ reduction million tonnes
Efficient tobacco barn	−63.25	6.03
Sewage biogas	−46.67	10.41
Efficient furnace	−44.91	44.50
Efficient lighting	−41.10	0.19
Industrial boilers	−31.38	10.17
Biogas for rural kitchens	−24.58	0.05

Figure 7.2 *Marginal reduction costs in US$ per tonne of CO$_2$ in Zimbabwe, gross financial costs (top figure), social costs alternative 1 (bottom figure)*

BOX 7.3 CHARACTERIZATION OF CASE EXAMPLES FOR BOTSWANA[1]

Case example	Project	Baseline case	Assumptions made in adjusting financial costs to get social costs
Road pavement	Pavement of 531km sandy roads that links Botswana to Zambia, Zimbabwe and Namibia	Sandy roads with similar traffic level, but with 50% higher fuel consumption for the same traffic	The social costs include the benefits of reduced SO_2, NO_x and particulate emissions Increased employment in the construction phase
Efficient lighting	Introduction of compact fluorescent lamps, 11 watt	Incandescent lamps, 60 watt Coal-fired electricity	The social costs include the benefits of reduced SO_2, NO_x and particulate emissions Health impacts implied by decreased coal production in the mines
Power factor correction	Installation of power factor correction	Coal-fired power plants without power factor correction Coal-fired electricity	The social costs include the benefits of reduced SO_2, NO_x and particulate emissions Health impacts implied by decreased coal production in the mines
Construction of pipeline for fuel oil	Pipeline of 400km is constructed to transport petroleum from Pretoria	Road and rail transport of petroleum	The social costs include the benefits of reduced SO_2, NO_x and particulate emissions Increased employment generated in the pipeline construction phase
Central PV plant	2 MW capacity of PV is established additional to the PV capacity that is assumed implemented in the baseline case	Coal-fired power production	The social costs include the benefits of reduced SO_2, NO_x and particulate emissions Health impacts implied by decreased coal production in the mines
Industrial boilers	Improved coal-fired boiler with 85% efficiency by installing economizer on traditional boilers	Existing coal-fired boiler with 79% efficiency	The social costs include the benefits of reduced SO_2, NO_x and particulate emissions Health impacts implied by decreased coal production in the mines
Vehicle inspection	Introduction of annual vehicle inspection service implying 10% efficiency improvement	Existing service levels	The social costs include the benefits of reduced SO_2, NO_x and particulate emissions

1 More details about the Botswana cases can be found in UNEP, 1999

from the power plants as well, and employment and health impacts related to the reduction of coal supplied from Botswana coal mines. The transportation projects similarly have indirect impacts on local air pollution, and in some cases also have significant positive local employment impacts due to large construction activities, as in the case of the road-pavement project.

Estimates of financial and social costs of the case projects for Botswana are shown in Table 7.5.

Table 7.5 *Financial and social costs of GHG emission-reduction projects in Botswana (all costs are in US$ per tonne CO_2 reduction, 10 per cent discount rate)*

	Gross financial costs	Net financial costs	Social costs alternative 1	Local air pollution impacts	Employment impacts	Social costs alternative 2	Coal-mining impacts
Vehicle inspection	1.7	1.6	1.2	−0.3	−	1.2	0.0
Efficient industrial boilers	5.0	−5.9	−57.6	−51.7	−	−221.8	−180.6
Paved roads	13.0	−101.2	−140.3	−28.3	−0.6	−140.3	0.0
Power factor correction	14.3	−7.9	−78.1	−70.2	−	−262.2	−184.0
Efficient lighting households	67.5	−113.7	−133.3	−19.65	−	−184.7	−51.4
Central PV	86.6	67.1	5.8	−60.2	−	−161.7	−184.2
Petroleum pipeline	181.1	125.6	79.6	−47.0	−	79.6	0.0

The Botswana projects as in to the case projects described for Zimbabwe, show a big difference between gross financial costs, net financial costs and social costs. Gross financial costs range from a cost of US$1.7 per tonne of CO_2 for the vehicle inspection case, up to a cost of US$181.1 per tonne of CO_2 reduction for the petroleum pipeline project. Net financial costs are in all cases lower than gross financial costs, in particular with a big difference for the road-pavement project and the efficient household lighting options, where the value of the fuel and electricity savings included in the net financial costs are very large. The road-pavement and the household-lighting projects have negative net financial costs of over US$100 per tonne of CO_2, which reflect that they are assumed to substitute very inefficient baseline activities. The industrial boiler and the power factor correction project also have negative costs.

Social costs are lower than financial costs for all the projects considered. The social costs in alternative 1, as the net financial costs, are negative for all projects except vehicle inspection and central PV that have small positive costs, and the petroleum pipeline project that has a high social cost of about US$80 per tonne of CO_2 reduction. In particular, the power correction factor, central PV and the efficient industrial boilers have large benefits on local air pollution, amounting to US$70 per tonne of CO_2, US$60 per tonne of CO_2, and US$52 per tonne of CO_2 respectively, because the projects substitute pollution-intensive coal consumption. These local air pollution benefits, however, are still not large enough to make the projects more cost-effective than the road-pavement and the lighting projects. The inclusion of coal-mining impacts, as reflected in the social cost alternative 2, once more make the industrial boiler project, the power factor correction project, the lighting project and the central PV project more cost-effective because they substitute coal consumption.

The cost-effectiveness ranking of the Botswana projects is shown in Table 7.6.

Table 7.6 *Ranking of GHG emission reduction projects according to cost-effectiveness (the project with the lowest cost per tonne of CO_2 reduction has been given a ranking value of one)*

	Gross financial costs	Net financial costs	Social costs alternative 1	Social costs alternative 2
Vehicle inspection	1	5	5	6
Efficient boilers	2	4	4	1
Paved roads electricity	3	1	1	4
Power factor correction	4	3	3	5
Efficient lighting households	5	2	2	2
Central PV for power production	6	6	6	3
Petroleum pipeline	7	7	7	7

The cost-effectiveness ranking of the projects change when the cost perspectives change from gross to net financial cost. The most attractive projects in the gross financial cost ranking are projects like vehicle inspection and efficient boilers that have relatively low capital costs. The ranking change in the net financial cost perspective, where projects like efficient lighting and road pavement move up to be the most cost-effective projects because they generate large energy savings. The vehicle-inspection project and the efficient boiler project, in contrast to the gross financial cost ranking, move down to be among the projects with a medium to low cost effectiveness. The project ranking in the social cost alternative 1 is almost the same as in the net financial costs alternative, but the inclusion of the coal-mining impacts in social cost alternative 2 makes a difference. The road-pavement and the vehicle-inspection projects here change to be low priority projects because they do not substitute coal-based activities.

Marginal emission reduction costs curves, based on gross financial costs and the social cost alternative 1, are shown in Figure 7.3. The cost curves depict net present values of the costs and GHG emission reductions of the projects calculated with a 10 per cent discount rate over the lifetime of the projects.

Seen from the social cost perspective, the cost curves suggest the paved roads as a very attractive option seen from the national perspective. This project in total is assessed to generate a domestic benefit of US$351 million in the social cost alternative 1, while the total gross financial cost of the project is an NPV of US$32.5 million. These gross financial costs must be considered as relatively low compared with many other GHG emission-reduction projects.[16]

The UNEP mitigation costing studies for Botswana also included, in addition to the assessment of direct costs, a qualitative discussion about the possible development impacts of the mitigation options (UNEP, 1999). The options were qualitatively ranked in relation to the following criteria:

- accordance with government policy;
- implementability;
- balance of trade;
- employment;
- social benefits;
- economic efficiency;
- benefits in other sectors;
- local environmental impacts.

These wider criteria were discussed in Chapter 5 and the options that came out with the highest ranking in this assessment were vehicle inspection and efficient lighting. Other options that achieved a relatively favourable ranking were road pavement, industrial boilers and power factor correction. The options with the lowest ranking in the qualitative assessment were central PV electricity and the petroleum pipeline. This project ranking shows up to be very similar to the ranking based on the social cost, alternative 1, being the only difference that vehicle inspection in the qualitative assessment has been given a higher priority than in the social cost assessment.

Mauritius Case Studies

The GHG emission-reduction projects assessed for Mauritius is a further development of a national mitigation costing study that has been conducted by the government of Mauritius and the UNEP Collaborating Centre on Energy and Environment as part of the GEF project Economics of Greenhouse Gas Limitations (Markandya and Boyd, 1999). The Mauritius country studies considered a number of GHG emission-reduction options for the energy and transport sectors.

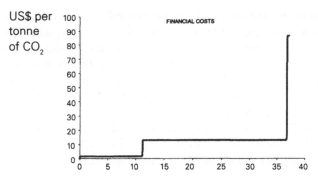

CO$_2$ reduction in millions of tonnes

Project	Gross financial costs US$ per tonne CO$_2$	CO$_2$ reduction million tonnes
Vehicle inspection	1.7	1.10
Efficient industrial boilers	5.0	0.04
Paved roads	13.0	2.50
Power factor correction	14.3	0.07
Efficient lighting	67.5	0.08
Central PV plant	86.6	0.04
Petroleum pipeline	181.1	0.40

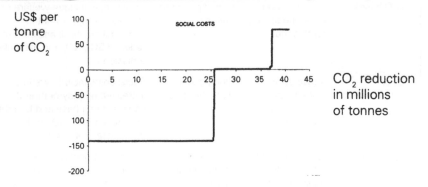

CO$_2$ reduction in millions of tonnes

Project	Social costs US$ per tonne CO$_2$	CO$_2$ reduction million tonnes
Paved roads	−140.3	2.50
Efficient lighting	−133.3	0.08
Power factor correction	−78.1	0.07
Efficient boilers	−57.6	0.04
Vehicle inspection	1.2	1.10
Central PV plant	5.8	0.04
Petroleum pipeline	79.6	0.40

Figure 7.3 *Marginal reduction costs in US$ per tonne of CO$_2$ in Botswana, 10 per cent discount rate, gross financial costs (top figure), social costs alternative 1 (bottom figure)*

The case projects selected for the current extended costing study include an LPG bus project and four renewable energy projects – namely, wind turbines, solar water-heaters, PV streetlights and bagasse power production. All the projects are expected to have domestic social benefits in the form of decreased local air pollution, and most of them are also expected to generate increased employment in the implementation phase. An overview of the GHG emission-reduction case projects is given in Box 7.4.

The projects considered for Mauritius all have relatively high costs compared with the costs of the case examples assessed for Zimbabwe, Botswana and Thailand. The gross financial costs in Mauritius range from US$406 per tonne

---

Box 7.4 Characterization of case examples for Mauritius[1]

Case example	Project	Baseline case	Assumptions made in adjusting financial costs to get social costs
Wind turbines	Installation of 30 MW wind farm	Fuel oil-based power production	Social costs include welfare of increased employment in the construction phase and benefits of reduced SO_2, NO_x and particulate emissions
LPG buses	LPG-powered buses dedicated for this fuel	Diesel-powered buses	Social costs include welfare of increased employment in construction phase and benefit of reduced SO_2, NO_x and particulate emissions
Solar water-heaters	Active solar water-heaters in domestic premises	Electric solar water-heaters	Social costs include welfare of increased employment in the construction phase and benefits of reduced SO_2, NO_x and particulate emissions
PV street lights	125 street lights identified for replacement are replaced with PV lights	Existing street light using fuel oil-based power	Social costs include welfare of reduced SO_2, NO_x and particulate emissions
Bagasse power production	Power production with a bagasse–coal mixture in addition to the capacity already planned in the baseline case	Fuel oil-based power production	The social costs include a negative employment value because the option is assumed to have no employment related to construction activities and and substitutes existing power production
			Social costs additionally include benefit of reduced SO_2, NO_x, and particulate emissions

1 More details about the project for Mauritius can be found in Markandya and Boyd (1999)

of CO_2 for the wind power project to as much as US$1839 per tonne of CO_2 for LPG buses. The net financial costs are lower than gross financial costs for all options with the largest decrease between gross and net financial cost seen for the bagasse power production project. The social cost components related to local air pollution benefits and employment are not large enough to get the CO_2 reduction price down to the level that is seen in the studies of other countries the range of the social costs being as high as between US$82 per tonne of CO_2 up to US$1299 per tonne of CO_2.

The main national social co-benefits of the case projects are seen in relation to local air pollution reduction for the LPG buses that will amount to US$76 per tonne of CO_2. The wind-power project and PV street-lighting projects are expected to have high benefits in the form of health improvements related to increased local employment in the construction phase. The bagasse power production project has lower employment than the baseline project and this implies that the social costs of this project are higher than the net financial costs.

The financial and social cost calculations for Mauritius are shown in Table 7.7. The cost-effectiveness ranking of the projects vary between the gross financial, the net financial and the social cost perspective, as shown in Table 7.8. The main difference is seen here for the bagasse power production project that is a third priority according to the gross financial cost perspective, but a first priority according to the net financial and social cost perspectives. This result is achieved because the bagasse project implies significant fuel savings in power production that are included in the net financial and the social costs, but not in the gross financial costs. No changes in cost-effectiveness ranking are seen for the other projects, and that is in particular a consequence of the relatively high capital costs of these projects.

Marginal emission-reduction costs curves based on gross financial costs and social costs are shown in Figure 7.4. The cost curves depict net present values of the costs and GHG emission savings of the projects calculated with a 10 per cent discount rate over the lifetime of the projects.

Table 7.7 *Financial and social costs of GHG emission-reduction projects in Mauritius (all costs are in US$ per tonne CO_2 reduction, 10 per cent discount rate)*

	Gross financial costs	Net financial costs	Social cost alternative 1	Local air pollution impact	Employment impact
Wind power	405.9	232.3	209.5	−3.3	−22.3
Solar water-heaters	448.7	312.0	278.3	−1.6	−32.1
Bagasse power production	561.0	79.7	82.0	−2.0	5.7
PV street lighting	1237.9	1068.3	971.7	−1.6	−76.4
LPG buses	1838.8	1370.6	1299.1	−76.0	−4.6

Table 7.8 *Ranking of GHG emission-reduction projects for Mauritius according to cost-effectiveness (the project with the lowest cost per tonne of CO$_2$ reduction has been given a ranking value of one)*

	Gross financial costs	Net financial costs	Social costs alternative 1
Wind power	1	2	2
Solar water-heaters	2	3	3
Bagasse power production	3	1	1
PV street lighting	4	4	4
LPG buses	5	5	5

According to the gross financial cost perspective, the wind power, solar water-heater and bagasse projects have relatively similar, but also high costs per unit of CO$_2$ reduction. The costs are more divergent across projects, when the social cost perspective is applied. The bagasse power project has significantly lower social costs of around US$80 per tonne CO$_2$ than the second most attractive project, wind power, and the third most attractive project, solar water-heaters, with social costs of around US$200 per tonne of CO$_2$ and around US$270 per tonne of CO$_2$ respectively.

For Mauritius it is worth noticing that there is a significant difference between cost estimates which are based on a 10 per cent discount rate used to calculate NPVs of costs and GHG emission reductions, and estimates based on a 10 per cent discount rate for the costs and a 3 per cent discount rate for the GHG emission reductions. A difference in GHG emission-reduction costs in particular is seen for the renewable energy projects included in the study, because these projects tend to have large up-front capital costs and generate annual GHG emission reductions over a long time horizon. The results of this sensitivity analysis are included in Appendix 7.3.

Thailand Case Studies

The GHG emission-reduction case projects that have been assessed for Thailand are selected on the basis of a national mitigation costing study, which have been carried out as part of the Asia Least-cost Greenhouse Gas Abatement Strategy (ALGAS, 1999b). The ALGAS study covers the energy sector, forestry and agriculture, and is based on sectoral modelling of abatement potentials and related costs. The GHG emission-reduction project case examples assessed in this extended analysis included 4 energy sector projects out of the 14 projects included in the energy sector study. The 4 projects were among the options that have been assessed to be possible to implement in a short- to medium-term time horizon (5–15 years). The projects assessed are efficient industrial motors, biomass cogeneration in industry, fuel-efficiency improvements for vehicles, and efficient woodfuel stoves for household cooking and for local industry. The

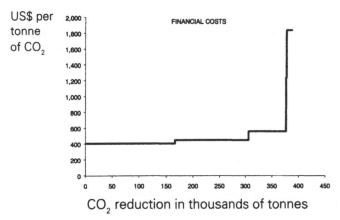

Project	Gross financial costs US$ per tonne CO_2	CO_2 reduction thousand tonnes
Wind power	405.9	0.170
Solar water-heaters	448.7	0.140
Bagasse power	561.0	0.070
PV street lighting	1237.9	0.001
LPG buses	1838.8	0.010

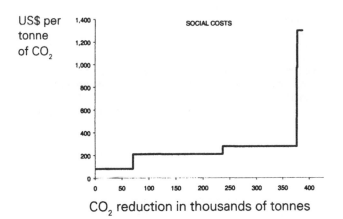

Project	Social costs US$ per tonne CO_2	CO_2 reduction thousand tonnes
Bagasse power	83.2	0.07
Wind power	201.8	0.17
Solar water-heaters	268.9	0.14
PV street lighting	949.3	0.001
LPG buses	1249.0	0.01

Figure 7.4 *Marginal reduction costs in US$ per ton of CO_2 in Mauritius, 10 per cent discount rate, gross financial costs (top figure), social costs alternative 1 (bottom figure)*

ALGAS study for Thailand also included a number of power production projects, which were an increased use of natural gas and the introduction of nuclear energy. These projects have not been included in the current case studies because such large-scale projects were expected to have major structural impacts on a power system that should be assessed in a sectoral modelling framework.

The GHG emission-reduction projects for Thailand all are expected to imply decreased local air pollution and this impact in particular is assessed to be large for the biomass cogeneration and the efficient woodfuel stoves. The efficient woodfuel stoves, in addition to the air pollution benefits, are also expected to generate time-saving benefits related to reduced woodfuel collection by the households. An overview of the project cases is given in Box 7.5. The financial and social costs of the case projects for Thailand are shown in Table 7.9.

Box 7.5 Characterization of case examples for Thailand[1]

Case example	Project	Baseline case	Assumptions made in adjusting financial costs to get social costs
Efficient industrial motors	Efficient motors with 91% efficiency are introduced in manufacture (assumed penetration of 10,000 units)	Existing motors with 86% efficiency Average electricity supply	The social costs include the benefits of reduced SO_2, NO_x and particulate emissions from decreased power production
Cogeneration based on biomass	Expansion of existing capacity in biomass-based cogeneration Installation of 25-bar steam turbines	Existing capacity in cogeneration Average electricity supply	Emissions of SO_2, NOx and particulates from cogeneration The social costs include the benefits of reduced SO_2, NO_x and particulate emissions in the biomass cogeneration compared with the substituted power production
Fuel efficiency improvement in transport	Retrofit of old vehicles to reduce gasoline consumption	Existing engines	The social costs include the benefits of reduced SO_2, NO_x and particulate emissions from decreased gasoline consumption
Efficient stoves to reduce fuelwood consumption	Penetration of efficient stoves that can save 60% of fuelwood consumption	Existing stoves Decreased deforestation	The social costs include a benefit of reduced SO_2, NO_x and particulate emissions Time savings due to decreased wood fuel collection are valued; the value assigned to one hour saved is US$0.5

1 More details about the Thai projects can be found in ALGAS, 1999b

The gross financial costs range from US$3 per tonne of CO_2 for the efficient woodfuel stoves to US$141 for the biomass cogeneration project as seen in Table 7.9. Projects like the woodfuel stoves and the industrial motors are, from this perspective, very cost-effective, reflecting their low capital costs. The net financial costs are significantly lower than the gross financial costs for the industrial motors, efficient vehicles and the biomass cogeneration option, reflecting that the value of the fuel and electricity savings achieved with these options are very large. Two of the projects – namely, the efficient motor and vehicle options – have negative net financial costs, because they imply fuel costs savings of over US$100 per tonne of CO_2 reduction.

The cost-effectiveness ranking of the options change, when the social cost perspective is applied. In particular the woodfuel stove project becomes attractive, which reflects that this project has been assessed to have large social benefits due to decreased time spent on woodfuel collection.[18] Another social benefit is reduced local air pollution that in particular has been assessed to generate a high value in the case of biomass cogeneration. The cost-effectiveness ranking of the projects in relation to the different cost perspectives is shown in Table 7.10.

A general conclusion on the ranking presented in Table 7.10. is that the industrial motor project, both in the gross financial, the net financial and the social cost-based ranking, is among the most cost-effective options. Another option that has a similar ranking in the three different cost perspectives is the efficient vehicle project that is assessed to have a relatively low-cost effectiveness.

Marginal emission-reduction costs curves based on gross financial and social costs are shown in Figure 7.5. The cost curves depict net present values of the costs and GHG emission savings of the projects calculated with a 10 per cent discount rate over the lifetime of the projects.

The two most significant case projects with regards to CO_2 reduction potential are the efficient vehicles and the biomass cogeneration project. These two

Table 7.9 *Financial and social costs of GHG emission-reduction projects in Thailand (all costs are in US$ per tonne CO_2 reduction, 10 per cent discount rate)*

	Gross financial costs	Net financial costs	Social cost alternative 1	Local air pollution impacts	Employment impacts
Efficient wood-fuel stoves	2.8	2.8	–145.3	–11.8	–136.4[1]
Efficient motors in industry	13.4	–95.7	–100.4	–4.7	–
Efficient vehicles	56.9	–74.7	–74.8	–0.5	–
Biomass cogeneration	141.4	54.6	17.8	–36.8	–

1 Time savings due to decreased time spent on woodfuel collection

Table 7.10 *Ranking of GHG emission-reduction projects according to cost-effectiveness (the project with the lowest cost per tonne of CO_2 reduction has been given a ranking value of one)*

	Gross financial costs	Net financial costs	Social costs alternative 1
Efficient wood fuel stoves	1	3	1
Efficient motors in industry	2	1	2
Efficient vehicles	3	4	3
Biomass cogeneration	4	2	4

projects will save respectively 40 million tonnes of CO_2 and 168 million tonnes of CO_2 (NPVs discounted with 10 per cent). The projects are relatively expensive seen from a gross financial cost perspective, with a cost of US$57 per tonne of CO_2 for the vehicle project and a cost of US$141 for the cogeneration project. However, there are large fuel savings benefits and social benefits connected to these projects, which will make them attractive when seen from a national Thai point of view.

Cross-cutting Conclusions on the Country Study Results

This assessment of case projects has identified a potential for low-cost GHG emission reductions in Zimbabwe, Botswana and Thailand from a financial as well as a social cost perspective. Those projects with the lowest costs on both criteria could be worth implementing seen from both a global cost-effectiveness perspective and a national development perspective. The study for Mauritius generally estimated higher GHG emission-reduction costs than for the other countries included, which both reflects the specific character of the projects included in this study and that the power system of Mauritius has a relatively low GHG emission intensity. The power system baseline case for Mauritius has been fuel oil-based production, while the power system baseline case has been coal-based production for Zimbabwe and Botswana, and a mix of coal and fuel oil-based power production for Thailand.

The study for Zimbabwe included a number of projects that have low gross and net financial costs and social costs. These include the efficient industrial boilers, the efficient tobacco barn project, the biogas for rural households project and the sewage biogas plant, all of which are assessed to have gross financial costs below US$16 per tonne of CO_2 equivalent reduction. Seen from a social cost perspective the costs of these three projects ranged from –US$25 to –US$1119 per tonne of CO_2 equivalent reduction. Two other projects – namely, efficient lighting and efficient furnaces in manufacture – had slightly higher gross financial costs of up to US$58 per tonne of CO_2 reduction, but were assessed to have social costs of –US$41 to –US$45 per tonne of CO_2 reduction.

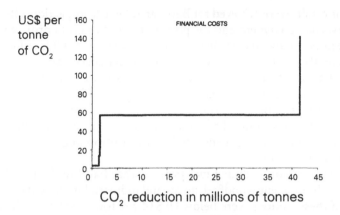

CO$_2$ reduction in millions of tonnes

Project	Gross financial costs US$ per tonne CO$_2$	CO$_2$ reduction million tonnes
Efficient woodfuel stoves	2.8	1.4
Efficient motors in industry	13.4	0.2
Efficient vehicles	56.9	40.0
Biomass cogeneration	141.4	168.3

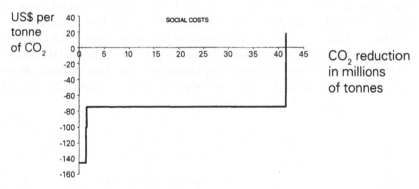

CO$_2$ reduction in millions of tonnes

Project	Social costs US$ per tonne CO$_2$	CO$_2$ reduction million tonnes
Efficient woodfuel stoves	−145.3	1.4
Efficient motors in industry	−100.4	0.2
Efficient vehicles	−74.8	40.0
Biomass cogeneration	17.8	168.3

Figure 7.5 *Marginal reduction costs in US$ per tonne of CO$_2$ in Thailand, 10 per cent discount rate, gross financial costs (top figure), social costs alternative 1 (bottom figure)*

Somehow, similar results were achieved for Botswana. Vehicle inspection, efficient industrial boilers, road pavement and the power factor correction project were assessed to have gross financial costs below US$15 per tonne of CO_2 reduction. The projects were also assessed to have very low social costs, ranging from – US$28 to –US$70 per tonne of CO_2 reduction, with the vehicle inspection project as an exception that was assessed to have a social cost of US$1.2 per tonne of CO_2 reduction. Another option that was assessed to have a low social cost was efficient lighting in the household. The low social costs of this option both reflect a high value of electricity savings generated by this option and positive side-impacts on local air pollution.

The study for Thailand included two projects that had low financial and social costs – namely, efficient woodfuel stoves and industrial motors. The gross financial costs of these options were below US$14 per tonne of CO_2 reduction and the social costs were below –US$100 per tonne of CO_2 reduction. Efficient vehicles and biomass cogeneration were assessed to have higher gross financial costs, but were assessed to have social costs below US$418 per tonne of CO_2 reduction. In particular, the value of fuel or electricity savings achieved by the GHG emission-reduction options were assessed to be large for the projects assessed for Thailand.

Against this background, it can be concluded more generally that the studies for Zimbabwe, Botswana and Thailand included a number of energy efficiency options for the manufacturing sector and the household and transport sectors that have been assessed to have relatively low gross financial costs. The net financial costs of the options that included benefits of fuel or electricity savings and the social costs similarly were very low for these options. A number of other projects assessed for these countries have higher gross financial costs than the aforementioned energy efficiency improvements, but these options change to be among the low-cost GHG emission-reduction options, when a net financial cost or a social cost perspective is applied.

The current project assessment includes in total 22 cases, all of which, except one, showed up to have lower social than financial costs (gross as well as net financial costs). This is the case because the projects were assessed to imply increased local air pollution and in some cases also local employment that had social benefits. The only project that was assessed to have higher social costs than net financial costs was bagasse power production in Mauritius which was expected to generate a net decrease in local employment that was larger than the local air pollution benefit associated with the option. The benefits of the side-impacts on local air pollution range between US$27 and US$36 per tonne of CO_2 reduction for most of the Zimbabwean cases, between US$28 and US$70 for most of the Botswana cases, between US$1.6 and US$3 for most of the Mauritius cases, and between US$0.5 and US$37 for the Thai cases. It should be recognized that the benefit estimates of avoided local air pollution are based on international 'benefit transfer', as pointed out in the section on the methodological framework. These international benefit transfers, as it was pointed out earlier, must be seen as an inaccurate estimate of specific national damages.

The case studies for Zimbabwe and Botswana have included an additional social cost alternative, where the side-impacts of increased coal production in local mines have been assessed. The side-impacts included occupational health impacts in the form of mortality and injuries. The coal-mining impacts were assessed to be very large for both countries, and the inclusion of these in particular, therefore, made projects that substituted coal-based activities very attractive. It must be recognized that the benefits implied for reduced coal-mining in this study did not include the eventual disbenefits of unemployment generated by reduced coal-mining activities. Employment impacts might well be very important national decision criteria in relation to the selection of projects that substitute coal-based activities.

The cost-effectiveness ranking of the GHG emission-reduction projects has shown up to be very different in the gross, net and social cost perspectives. A number of the projects that were the most cost-effective in a gross financial cost ranking moved down on the priority list when a net financial and a social cost perspective were applied. In particular, this reflects that some projects like efficient lighting, energy efficiency improvements in the industrial sector, and a number of fuel-saving options in the transport sector has significant energy-saving benefits and local air pollution benefits. This in turn means that projects that would seem to be the most attractive from a project implementing agents perspective (international or national agents) in many cases will not be the same as the projects that generate large local development benefits. The selection of GHG emission projects in this way is critically dependent on the perspective and the objectives reflected in the cost-effectiveness analysis.

GENERAL CONCLUSIONS

The current study has assessed the financial and social costs of 22 GHG emission-reduction case projects for Zimbabwe, Botswana, Mauritius, and Thailand, selected out of national mitigation costing studies conducted as part of international country study programmes by the UNEP and the Asian Development Bank, the Global Environment Facility (GEF) and the UNDP.[19] The national mitigation cost studies estimated the direct costs of projects and sectoral strategies. In the current study the assessed costs were extended to include the monetary value of a number of social indicators reflecting local air pollution impacts, employment generation, and the health impacts in the coal-mining sector. The study has been based partly on specific detailed national assessments and partly on approximated data on air pollution damages and health impacts. The study is therefore not providing an accurate estimate of the GHG emission-reduction costs of the projects, but should rather be seen as useful in providing a number of general insights about differences in the cost-effectiveness of the projects, given specific financial and social cost concepts. Indeed, one of the recommendations is that primary work is needed in the developing countries to arrive at more accurate estimates of the costs/benefits arising from these indirect/secondary impacts of the mitigation programmes.

The social cost concepts defined and applied in this project assessment are an example of how a number of local sustainable development impacts of CDM projects can be assessed. It can very well be argued that the sustainable development impacts of CDM projects will range beyond the actual employment, local air pollution and coal-mining impact indicators included in this study, and an assessment of CDM projects therefore might include more indicators that can be assessed quantitatively as well as qualitatively. Despite the limited number of indicators assessed for the GHG emission-reduction projects considered in this study, it is worth noticing that the magnitude of the GHG emission-reduction costs, as well as the cost-effectiveness ranking of projects, are closely related to the financial and social cost accounting frameworks. A gross financial cost perspective that primarily reflects the capital costs of the options per unit of GHG emission-reduction results in a cost-effectiveness ranking of the projects that is significantly different from the ranking that comes out of a financial and social cost-based ranking. If the gross financial costs are taken as the perspective of project donors, the actual projects initiated from that perspective will not be the same as the ones that would emerge from a greater emphasis on local benefits in the form of energy savings and social improvements.

Another general conclusion is that the net financial costs as well as the social costs in all cases are lower than the gross financial costs. This implies that a project host country receives local benefits if an international donor is willing to supply the gross financial cost of implementing the project. The actual magnitude of the benefits in the form of energy savings and social benefits are, however, very site specific. The energy savings depend on the efficiency of the baseline energy system, while the social benefits depend on a broad range of site- and time-specific issues such as employment, local air pollution and health-related issues. In particular, the net financial costs and the social costs will vary from case to case for similar categories of GHG emission-reduction projects.

NOTES

1   In specific project implementation mechanisms like the Clean Development Mechanism (CDM), implementation costs will also include the costs of administration, monitoring and certification of carbon reductions, which have not been included in this assessment.
2   The VOSL approach has excited much controversy and some policy-makers remain unconvinced as to its validity. The other approach, which is to value individuals in terms of their economic productivity, is, however, even more fraught with logical and ethical contradictions.
3   Particulate emissions have not been included in the case studies for Zimbabwe and Botswana.
4   Based on Markandya (1998), Tables 9 and 10, but adjusted with a recent update of PPP GDP (WDI, 1998).
5   The damage values for Mauritius have been adjusted by a scaling factor that reflects differences in population density between Mauritius and the European sites (UK and Germany) that have been used to establish the ExterneE data. The UK Mauritius scaling factor is 1.95 and the Germany Mauritius scaling factor is 1.37.
6   The baseline case is for the power systems a sectoral scenario for power supply over a 20-year timeframe in the national context.
7   The time dependence of the climate change damage 'function' reflects that the total stock of atmospheric greenhouse gases varies with time and the 'damage' of emitting one unit of emission depends on this stock.
8   More detailed arguments for this approach are provided by Halsnaes, Callaway and Meyer (1998).

9   Climate change damages are a global common good and avoided damages therefore have been considered as a social benefit.
10   CH_4 emissions are converted to CO_2 equivalent emissions with the use of global warming potential (GWP) values.
11   The baseline case for this project is local woodfuel collection, which is assumed to have zero financial costs.
12   The high sulphur content of Zimbabwean coal (2.5 per cent) generates high local air pollution levels.
13   The fuel expenditure savings related to reduced coal consumption are assumed to be relatively low due to a plentiful and inexpensive domestic coal supply.
14   Supplemented with a reforestation option and a no-tillage option for agriculture.
15   More details about the Botswana cases can be found in UNEP, 1999.
16   It could be argued that the GHG emission reductions achieved by the road-pavement project potentially would have a risk of being offset by traffic increase induced by the improved road facilities. Such a traffic increase, however, could also be seen as a positive side-impact on development which, similarly to economic development in general, will generate necessarily increased traffic and higher emissions.
17   More details about the project for Mauritius can be found in Markandya and Boyd, 1999.
18   Time calculation is based on the assumption that one hour has a value of US$0.5 and 10 kg of dry matter can be collected in one hour.
19   For financial costs, two measures are used: gross financial costs, which include all monetary outlays associated with the project, and net financial costs, which subtract from the former any financial savings that the project generates to parties other than those undertaking the project. For social costs, two measures are also used in the case of Botswana. One excludes any occupational health impacts on coal-miners and other includes such impacts.

REFERENCES

ALGAS (1999a) *Asian Least-Cost Greenhouse Gas Abatement Strategy*, Summary Report, Asian Development Bank, Manila, The Philippines
ALGAS (1999b) *Asian Least-Cost Greenhouse Gas Abatement Strategy, Thailand*, Asian Development Bank, Manila, The Philippines
Department of Mines Botswana (2000) *Tables on local air concentrations of sulphur and particulates* (photocopy)
Dessus, S and O'Connor, D (1999) *Climate Policy without Tears: CGE-Based Ancillary Benefits Estimates for Chile*, OECD, Paris (draft)
ExternE (1995) *External Costs of Energy: Volume 2 Methodology*, EC, Luxembourg
ExternE (1999) *External Costs of Energy: Volumes 1–10*, EC, Luxembourg
Halsnaes, K (1996) 'The economics of climate change mitigation in developing countries', *Energy Policy*, Special Issue, Vol 24, No 10/11
Halsnaes, K (1997) 'Cross-sectoral assessment of mitigation options', *Energy Policy*, Vol 25, No 2
Halsnaes, K, Callaway, J M and Meyer, H (1998) *Methodological Guidelines*, Economics of Greenhouse Gas Limitations, Main Report, UNEP Collaborating Centre on Energy and Environment, Risø National Laboratory, Denmark
Holland, M et al (1998) *Global Trends in the Health Risks of Coal Mining* (in preparation)
IPCC (1996) *Climate Change 1995. Economic and Social Dimensions of Climate Change. Scientific-Technical Analysis*, contribution of Working Group III to the Second Assessment Report of the Intergovernmental Panel on Climate Change, Cambridge University Press
Liu, J, Hammitt, J K and Liu, J L (1997) 'Estimated hedonic wage function and value of life in a developing country', *Economics Letters*, Vol 57, No 3, pp353–358

Markandya, A (1996) 'Economic Valuation' in: *ExternE: Externalities of Energy – Volume 2 – Methodology,* European Comission, DGXII, Brussels

Markandya, A (1998) *The indirect costs and benefits of greenhouse gas limitations: Mauritius Case Study,* Economics of Greenhouse Gas Limitations, Handbook Reports, UNEP Collaborating Centre on Energy and Environment, Risø National Laboratory, Denmark

Markandya, A (2001) 'The Valuation of Health Impacts in Developing Countries', in R. Serôa Da Motta (ed) *Environmental Economics and Policy-Making in Developing Countries: Current Issues,* Edward Elgar, pp82–106

Markandya, A and Boyd, R (1999) *The indirect costs and benefits of greenhouse gas limitations: Mauritius Case Study,* Economics of Greenhouse Gas Limitations, Handbook Reports, UNEP Collaborating Centre on Energy and Environment, Risø National Laboratory, Denmark

Pearce, D W (1996) 'Economic valuation and health damages from air pollution in the developing world', Energy Policy, Vol 24, No 7, pp627–630

Portney, P R and Weyant, J P (eds) (1999) *Discounting and Intergenerational Equity,* Resources for the Future, Washington, DC

Rowe, R, Bernow, S, Chestnut, L and Ray, D (1996) *New York State Externality Study,* Oceana Publications Inc., Dobbs Ferry, NY

Shah, J, Nagpal, T and Brandon, C (eds) (1997) *Urban Air Quality Management Strategy in Asia: Guidebook,* The World Bank, Washington, DC

Southern Centre (1993) *UNEP Greenhouse Gas Abatement Costing Studies, Zimbabwe Country Study, Phase Two,* Southern Centre for Energy and Environment, Department of Energy, Ministry of Transport and Energy, and Risø National Laboratory, Denmark

Southern Centre (1995) *UNEP Greenhouse Gas Abatement Costing Studies, Zimbabwe Country Study, Phase Three,* Southern Centre for Energy and Environment, Department of Energy, Ministry of Transport and Energy, and Risø National Laboratory, Denmark (photocopy)

Thayer, M A et al (1994) *The Air Quality Valuation Model,* Regional Economic Research Inc. and TRC Environmental Consultants

UNEP (1994) *UNEP Greenhouse Gas Abatement Costing Studies. Part One: Main Report,* UNEP Collaborating Centre on Energy and Environment, Risø National Laboratory, Denmark

UNEP (1999) *Economics of Greenhouse Gas Limitations. Country Study Series Botswana,* UNEP Collaborating Centre on Energy and Environment, Risø National Laboratory, Denmark

Watts, W (1999) *Discounting and Sustainability,* The European Comission, DGII, Brussels (photocopy)

WDI (1998) *World Development Indicators,* The World Bank, Washington, DC

Zilahy, G, Nemcsins, Z S, Szesler, A, Urge-Vorsatz, D, Markandya, A and Hunt, A (2000) *The indirect costs and benefits of greenhouse gas limitations: Hungary Case Study,* Economics of Greenhouse Gas Limitations. Handbook Reports. UNEP Collaborating Centre on Energy and Environment, Risø National Laboratory, Denmark

APPENDIX 7.1

Note on the Inclusion of Occupational Health Risks in Analysis of GHG Emission-reduction Projects

The analysis of projects for the reduction of GHGs has to date included, in some cases, the impacts of increased employment. The analysis of other secondary benefits or costs of GHG reduction projects, including impacts on the level of the health risk faced by workers, has not yet been included. This note attempts to draw together the literature on the valuation of the risks involved in coal-mining, drawing largely on the work of ExternE (1995 and 1999) and Holland et al (1998).

It should be noted that the extent to which occupational health risks should be included in analysis of the costs of energy generation depends largely on the extent to which the health risks are internalized, through payment to individuals to compensate for the risks they may face. In Europe there is some debate on the extent to which these health risks are internalized, although it is unlikely that the labour market in developing countries such as China, Botswana and Zimbabwe would internalize these impacts fully. The internalization process requires a good degree of knowledge of the health risks on the part of those facing the risk. It is our contention that the health risks in developing countries are minimally internalized, given the unemployment rates and the levels of education in developing countries, particularly in Africa.

Classification of Health Risks Faced by Miners

The health risks on which we have been able to focus fall largely into four categories:

- Risk of mortality from accident.
- Risk of serious injury.
- Risk of minor injury.
- Risk of pneumoconiosis.

This is by no means an exhaustive list of the risks faced by miners. Data requirements, however, have prevented a more extensive evaluation of the negative health risks faced by miners, although those not valued are largely the lesser morbidity impacts (phlegm, incidence of cough and others), which will be of little consequence to the overall analysis.

The impact pathways for mortality and injury are obvious. However, the impact of pneumoconiosis on health status is open to some debate. ExternE (1999) highlights the fact that pneumoconiosis in and of itself may not necessarily impair the mobility of an individual. The impact that ExternE identified from the incidence of this occupational disease was the increased risk of developing progressive massive fibrosis (PMF) which may lead to disablement or death. The approach of ExternE was to estimate the risk of developing PMF directly and value only the mortality impacts, taking an estimate of 2 per cent as the proportion of miners who get PMF who subsequently die of pneumoconiosis. The analysis conducted here will draw on this work.

Estimation of the Health Risks Faced by Miners

Holland et al (1998) present data on some of the major health risks faced by miners – notably, the risk of death by accident, risk of injury (serious and non-serious) and risk of pneumoconiosis – in 27 countries. Of the three countries of particular interest in the estimation of decreased health risk due to reduced coal production, only Chinese data is available. For Botswana and Zimbabwe, estimates from South Africa will be used as a lower bound of damages. Table 7.11 provides statistics for accident and occupational disease rates in selected countries.

Table 7.11 *Death and accident rates in selected countries*

Country	Fatality	Accident Major injury	Minor injury	Disease Pneumoconiosis (2)
South Africa	0.33	3.00	n/a	n/a
China (1)	4.90	n/a	n/a	0.09
USA	0.11	14.00	3.70	0.35
UK	0.11	5.40	35.00	0.05
Brazil	0.39	n/a	n/a	n/a
India	0.75	2.80	n/a	0.22

Notes:
(1) China is a composite of state and private mines. State mines have a far lower accident rate (see text).
(2) Pneumoconiosis cases are considered deaths. The China figure, based on ExternE (1999), estimates that 2 per cent of pneumonoconiosis cases result in death through PMF
Source: Adapted from Holland et al (1998)

From Table 7.11 it can be seen that the incidence of fatal accidents in China is substantially higher per million tonnes (Mt) of coal produced than that in the other countries featured. However, the Chinese figures are made up of a composite of state and private mines, with accident rates in state mines of 1.2 deaths/ Mt of coal produced, while private mines have a substantially worse safety record (8.4 deaths/Mt). China is by no means the worst country in terms of the fatality rate, Holland et al (1998) suggest that the death rates per Mt of coal are worse in Romania (10 deaths/Mt), Pakistan (30 deaths/Mt) and Turkey (119 deaths/ Mt). The range of values is quite broad, which emphasizes the need for specific country studies on occupational health risks in mines. South Africa ranked sixth best of 23 nations in Holland et al (1998) and hence should be considered a reasonable lower bound for countries in Africa.

Deaths due to pneumoconiosis are reported in 17 of the 23 countries. In Table 7.11 the figure reported for China (0.09 deaths/Mt of coal produced) is based on figures of cases of pneumoconiosis reported in Holland et al (1998), multiplied by a 2 per cent probability of death resultant from the development of PMF and pneumoconiosis. It is interesting that the estimate places China below the USA for deaths caused by pneumoconiosis per Mt of coal produced.

ExternE (1999) recommends the use of average data on risks of occupational health over a five-year period in recent history. The use of the average ensures that the effects of major accidents are smoothed and the application of recent data is necessary to reflect changes in the use of technology in the mines. The calculation of risk from occupational statistics necessarily makes the assumption that marginal and average accident rates are identical. This is not necessarily the case and decreases in mining activity may result in the closure of less safe mines.

Valuation of Occupational Health Impacts

ExternE (1995 and 1999) provides a framework for the valuation of occupational health hazards, using the impact pathway methodology. This framework will be applied for the case in Botswana, Zimbabwe and China to value the reduction in health risks due to reductions in coal-mining. As mentioned above, data limitations have restricted the application of this methodology to the major impacts: those of deaths from accident, major and minor injuries resultant from coal-mining and mortality resulting from increases in the cases of PMF. Other endpoints may be significant, although for illustrative purposes these major endpoints provide a good preliminary estimate of the value of reducing coal-mining from the point of view of occupational health (note that other benefits, such as reduced emissions from coal-fired plants, are considered elsewhere).

The ExternE (1999) values of reduced accident levels were reported separately from other costs, as internalization of some of these impacts may have been taking place in Europe. However, the extent to which occupational risk is contained in wage determination in developing countries is certainly less than in Europe, owing to factors such as the structure of the labour market and the bargaining power of employees. Liu, Hammitt and Liu (1997) suggest that compensating wage differentials were present in Taiwan, and propose a value of life of 1990 US$413,000, which can be compared to the Markandya (1998) estimate for China of 1995 US$388,000 to US$1.78 million. This may suggest that wage differentials do internalize some of the risks in developing countries, although the extent to which Taiwan can be considered comparable to China or Africa is debatable. This is an area that requires further investigation.

Mortality endpoints will be value-based on 1995 US VOSL calculations, adjusted to Botswana, Zimbabwe and China using PPP GNP values (Markandya, 1998; WDI, 1998). The values for these countries are reproduced in Table 7.12. Accident costs, following ExternE (1999), will be based on transfers of UK WTP estimates to avoid serious and slight accidents. The UK 1995 ECU values are presented in Table 7.13. These were adjusted using the same PPP transfer, compared to 1995 values of UK PPP GDP. The results of this transfer for Botswana and Zimbabwe are shown in Tables 7.14 and 7.15. Note that the values were only applied for an elasticity of 1.

Results

The results suggest that there are significant health benefits to be made from the switch from coal to other forms of energy production, owing to occupational health differences. These remain to be adjusted for the following:

- Changes in employment rates: where unemployment is the result of a switch from coal to other forms of energy production as a result of GHG reduction programmes, an adjustment of 4.5 deaths per 1000 unemployed has to be made (Markandya, 1998).
- Changes in income distribution across the countries. This may be an important factor in the analysis of different GHG reduction projects.

Table 7.12 *VOSL estimates*

Country	VOSL estimates (US$1995)
Botswana	1085
China	388
UK	2842
United States	4000
Zimbabwe	315

Source: Markandya (1998) and WDI (1998)

Table 7.13 *Accident costs: UK estimates*

Casualty severity	Human costs	Medical and support	Total
Serious	86,550	8500	95,050
Slight	6335	635	6970
Average	25,330	1795	27,125

Source: ExternE (1999)

Table 7.14 *Adjusted accident values: Botswana*

Casualty severity	Human costs	Medical and support	Total
Serious	33,079	3248	36,388
Slight	2421	243	2665
Average	9681	687	10,367

Source: Author's calculations based on ExternE (1999) and Markandya (1998)

Table 7.15 *Adjusted accident values: Zimbabwe*

Casualty severity	Human costs	Medical and support	Total
Serious	9610	944	10,554
Slight	703	71	774
Average	2812	199	3012

Source: Author's calculations based on ExternE (1999) and Markandya (1998)

APPENDIX 7.2

Overview of Options Included in the Direct Cost Studies for Zimbabwe, Botswana, Mauritius and Thailand

Table 7.16 *Main categories of GHG emission-reduction options included in UNEP studies for Zimbabwe*

Industry
Efficient boilers
General savings
Efficient motors
Efficient furnaces NH_3 from coal
Efficient tobacco barns

Agriculture
Efficient motors
Zero tillage
Methane from ruminants
Solar PV

Residential
Prepayment meters
Geyser timers
Solar water-heaters
Biogas digesters
Solar geysers
Efficient lighting

Transportation
Ethanol for blend
Power generation
Hydro power
Power factor correction
Central PV

Solid fuels
Coke oven gas in power generation
Methane from coal-mines

Forestry
Commercial forestry
Rural afforestation

Waste
Landfill gas
Urban sewage plant

Table 7.17 *Main categories of GHG reduction options in UNEP study for Botswana*

Road freight to rail
Efficient lighting
Tillage
Pipeline
Prepayment meters
Geyser time switches
Solar home systems
Power factor correction
Efficient boilers
Fuel pricing
Biogas from landfills
Vehicle inspection and efficient motors
Solar geysers
PV electricity
Biogas rural households
PV waterpumps
Re-forestation
Diesel to electric rail

Table 7.18 *Main categories of energy sector GHG emission-reduction options included in the ALGAS study for Thailand*

Lighting in the residential sector
Lighting in the commercial sector
Fuel efficiency improvements in transport
Cooling systems in the commercial sector
Industrial boilers
Refrigerators in the residential sector
Efficient motors in industry
Cogeneration in industry
Coal to natural-gas power production
Air-conditioning in the residential sector
Coal to natural gas in power production
Gas to nuclear power production
More nuclear power than previous options

Table 7.19 *GHG emission-reduction options included in the UNEP study for Mauritius*

PV street lights
Bagasse power production
Electricity price increase
Wind turbines
Solar water-heaters
LPG buses

APPENDIX 7.3

Sensitivity of the Options According to Discount Rates Used in Calculating NPVs of Costs and GHG Emission Reductions

Table 7.20 *Zimbabwe, cost calculations based on a 10 per cent discount rate for costs and GHG emission reductions, versus cost calculation based on a 10 per cent discount rate for costs and a 3 per cent discount rate for GHG emission reductions*

	Gross financial costs 10%/10%	Gross financial costs 10%/3%	Net financial costs 10%/10%	Net financial costs 10%/3%	Social costs alternative 1 10%/10%	Social costs alternative 1 10%/3%	Social costs alternative 2 10%/10%	Social costs alternative 2 10%/3%
Efficient tobacco barn	3.8	1.8	−5.9	−2.8	−63.3	−30.4	−107.5	−51.7
Biogas for rural kitchens	5.3	3.6	5.3	3.6	−24.6	−16.7	−24.6	−16.7
Sewage biogas for electricity	12.4	7.0	−19.6	11.0	−46.7	−26.2	−119.4	−67.0
Efficient industrial boilers	16.1	7.1	−3.1	−1.4	−31.4	−13.9	−109.0	−48.3
Efficient lighting	44.3	34.4	−6.8	−5.3	−41.1	−32.1	−105.9	−82.6
Efficient furnaces in manufacture	57.8	41.6	−8.1	−5.8	−44.9	−32.3	−128.8	−92.8

Table 7.21 *Botswana, cost calculations based on a 10 per cent discount rate for costs and GHG emission reductions, versus cost calculation based on a 10 per cent discount rate for costs and a 3 per cent discount rate for GHG emission-reductions*

	Gross financial costs 10%/10%	Gross financial costs 10%/3%	Net financial costs 10%/10%	Net financial costs 10%/3%	Social costs alternative 1 10%/10%	Social costs alternative 1 10%/3%	Social costs alternative 2 10%/10%	Social costs alternative 2 10%/3%
Vehicle inspection	1.7	1.2	1.6	1.2	1.2	0.8	1.2	0.8
Efficient industrial boilers	5.0	2.9	−5.9	−3.4	−57.6	−33.0	−221.8	−16.7
Paved roads	13.0	7.4	−101.2	−57.9	−140.3	−80.3	140.3	80.3
Power factor correction	14.3	7.1	−7.9	−1.4	−78.1	−13.9	−262.2	−48.3
Efficient lighting	67.5	54.3	−113.7	−91.4	−133.3	−107.2	−184.7	−148.5
Central PV	86.6	49.6	67.1	38.4	5.8	3.3	−161.7	−92.5
Petroleum pipeline	181.1	75.8	125.6	52.6	79.6	33.3	79.6	33.3

Table 7.22 *Mauritius, cost calculations based on a 10 per cent discount rate for costs and GHG emission reductions, versus cost calculation based on a 10 per cent discount rate for costs and a 3 per cent discount rate for GHG emission reductions*

	Gross financial costs 10%/10%	Gross financial costs 10%/3%	Net financial costs 10%/10%	Net financial costs 10%/3%	Social costs alternative 1 10%/10%	Social costs alternative 1 10%/3%
Wind power	405.9	243.3	232.3	139.2	209.5	125.6
Solar water-heaters	448.7	219.6	312.0	152.7	278.3	136.2
Bagasse power production	561.0	323.8	79.7	46.0	82.0	47.3
PV street lights	1237.9	719.7	1068.3	621.1	971.7	564.9
LPG buses	1838.8	875.6	1370.6	652.7	1299.1	618.6

Table 7.23 *Thailand, cost calculations based on a 10 per cent discount rate for costs and GHG emission reductions, versus cost calculation based on a 10 per cent discount rate for costs and a 3 per cent discount rate for GHG emission reductions*

	Gross financial costs 10%/10%	Gross financial costs 10%/3%	Net financial costs 10%/10%	Net financial costs 10%/3%	Social costs alternative 1 10%/10%	Social costs alternative 1 10%/3%
Efficient wood fuel stoves	2.8	2.3	2.8	2.3	−145.3	−120.3
Efficient motors in industry	13.4	9.6	−95.7	−68.9	−100.4	−72.3
Efficient vehicles	56.9	323.8	−74.7	46.0	−74.8	47.3
Biogas cogeneration	141.4	45.5	54.6	17.6	17.8	5.7

Chapter 8

The CDM and Sustainable Development: Case Studies from Brazil and India

Ronaldo Serôa da Motta, Leena Srivastava and Anil Markandya

INTRODUCTION

So far the methods of assessment for climate change mitigation that have been covered are cost-effectiveness analysis, cost-benefit analysis and multicriteria analysis, with examples of each being provided from actual case studies. In this chapter we look at two different approaches, one for Brazil and the other for India, both very important countries for the climate change issue generally and the developing country mitigation options debate in particular. The Brazilian study adopts an informal assessment approach. Impacts are identified but not quantified in money terms. Some indicators are provided and evaluated informally. The options are discussed in some depth and from a discursive assessment some priorities are identified. The end result is rich in detail and valuable to policy-makers, but it does not provide a formal ranking. The Indian study takes an approach that is akin to the multicriteria method. It gives rights to different outcomes, with the weights derived from consultations with government officials and experts. The extension to a standard multicriteria approach comes from methods used to convert non-quantified impacts into a scale that allows them to be compared across projects.

It is interesting to compare the results of these two very different approaches, not only with each other but with the other methods discussed in earlier chapters. This is done in the last section of the chapter.

CASE STUDY FOR BRAZIL[1]

As noted above, the Brazil study takes an informal approach in determining priorities. It was, however, only partly motivated by a desire to establish a ranking of priorities. Its other objective was to analyse the role of the market in a policy mechanism like the Clean Development Mechanism (CDM) of the Kyoto Protocol (UNFCCC, 1997); to help understand if a 'good' market will be created within a CDM institutional framework. Can CDM minimize global greenhouse gas (GHG) control costs and, at the same time, maximize each country's welfare? If not, which kind of intervention, or government regulation, will be needed which does not jeopardize incentives for cooperation?

The link with mitigation project appraisal comes from the fact that the efficiency of any CDM approach depends on what kinds of projects it ends up generating and how they compare with each other. The methodology is, therefore, an attempt to identify an expected ranking of investments which could come out through CDM, taking into account market forces interactions. The relative importance of each selected option is then made, when secondary benefits, which are not captured by private investors, are included. Based on that, it makes propositions of specific regulation options which a country may undertake to make CDM a global cost-effective mechanism compatible with the capture of local welfare gains.

HOW SHOULD SECONDARY IMPACTS OF MITIGATION PROJECTS BE EVALUATED?

Before examining the mitigation options for Brazil in detail, a case needs to be made for the non-monetary approach taken to evaluate the options. To date most studies that have attempted a serious assessment of the mitigation options in a broader context have, in fact, taken the monetary route, focusing mainly on energy-related projects and their environmental side-benefits related to local air pollution[2] (the so-called secondary benefits or secondary impacts). If, however, a wider assessment is to be achieved, account needs to be taken of factors that are not amenable to monetization, such as poverty alleviation and development linkages. Although these kinds of benefits are fully recognized in the literature, they have never been assessed in conjunction with each response option. This means that at least some of the impacts have to be assessed in a non-monetary way. The reason to avoid monetization for the impacts that *can* be estimated in monetary terms is partly that such estimates are controversial and partly that they are subject to great uncertainties. The establishment of monetary estimates requires the identification of economic relationships based on models, which quite often are biased and very sensitive to data quality. Secondly, the valuation itself, particularly in relation to environmental benefits, is extremely sensitive to the methodology chosen.

To illustrate this, one may take the literature on secondary benefits related to energy projects, as presented, for example, in Heintz and Tol (1996) and Ekins (1996). These are well-elaborated studies and identify very important issues on the subject. Nevertheless, secondary benefits in these studies are measured only as the health benefits of reduced urban air pollution concentrations. Moreover, the estimates are based on dose-response functions from developed countries and not on specific national data, because these data have not been available. This limited scope of the benefit estimates reveals important methodological and data biases. For example, dose-response functions can only be transferred to other places if air pollution concentrations and other socioeconomic variables are adjusted. Additionally, as the literature on cost-benefit analysis points out, there is a great divergence between willingness to pay (WTP) and avoided costs approaches when one is measuring health benefits.

Other limitations could always be brought about when one is analysing welfare measurements, as the authors of these studies also recognize. A reason not to go down the monetary valuation route is the confusion caused by adding together money values based on actual financial flows and those derived from a valuation, where there are no such flows. It is better, in our view, to reserve money values for the case where there are actual flows and to present other impacts separately.

Other issues that should be recognized in the assessment of secondary impacts of CDM projects include impacts on income distribution and employment generation. For example, reducing air pollution in big cities in Brazil is highly desirable because a large number of individuals living in these cities will benefit from the improved air quality, but it will be less socially important if improvements take place in less populated areas. Likewise, employment opportunities are a good thing, but especially desirable when employment increases for individuals belonging to the low-income group segment. These qualitative approaches, which are rather subjective, may disappoint formal economists, but they need to be considered in the analysis in order to avoid the previously stated controversial issues related to the monetary valuation of secondary impacts.

Based on these arguments, our assessment has been based on qualitative indicators reflecting on a number of non-monetary quantitative indicators. The result is then presented as a multidimensional qualitative information about the relevance, sign and dimension of specific secondary benefits, and no attempt is made to compare benefits across response options. The secondary impacts are classified into three categories: environmental, equity and developmental impacts. The analysis is limited to the assessment of secondary benefits of each individual option versus its private returns. Although this approach is used in the current study, this should not be understood as a general statement against monetary valuation as such. Further research and data gathering in order to improve the quality and accuracy of policy analysis is strongly recommended in order to establish a better understanding of how an innovative environmental policy mechanism like the CDM will work in the future.

Finally, it is to be emphasized that the current study does not present a social ranking of the Brazilian response options to be compared with a ranking that could emerge if the private sector alone were to have its priorities in a CDM framework. The analysis first of all is dedicated to show the existence of non-market impacts (positive as well as negative) of specific CDM projects.

To sum up, an 'indicative' or informal evaluation has been used in this study because this has been considered to be more politically understood than a partial and ad hoc monetary valuation of secondary policy impacts.

Climate Change Response Options in Brazil

Despite GHG emission-reduction options being numerous, the current study for simplification only considers two broad categories – namely, energy and forestry-related projects. Energy projects include fossil-fuel substitution measures and energy efficiency improvements. When a project implies increased biomass use, energy projects imply increased carbon sequestration in the plant growth phases. Forestry projects work through carbon sequestration. Atmospheric carbon is captured during the biomass growth process and can reduce net GHG emissions if deforestation is curbed.[3]

The Brazilian electricity structure was based historically on the expansion of hydroelectricity. Consequently, GHG (mainly CO_2) emissions, as a consequence of energy production, were traditionally low. However, Brazil now faces a conjunction of three important factors: (1) the hydroelectric potential in many parts of the country is very close to being fully utilized in the range of economically viable options, particularly near the most important consumption centres; (2) the relatively high (and highly variable) cost of oil products; and (3) the availability of natural gas in bordering countries (Bolivia and Argentina). Combined, these elements indicate that a considerable share of the new energy supply, required to satisfy the increasing needs of development, will be provided by thermoelectric options. This represents a structural change in the Brazilian energy system, and a direct consequence of this transformation will be increasing GHG emissions.

Ethanol as fuel for vehicles, once largely induced in the country, can no longer compete with gasoline and diesel on a production cost basis, although this critically depends on the price of crude oil, over a time horizon of at least 10–15 years.

There is no doubt that in the business-as-usual scenario, sound economic policy cannot discount fossil fuel options which would keep energy supply at pace with demand, particularly if one considers the country's needs of high growth vis-à-vis capital constraints and international competitiveness requirements. On the other hand, the continental dimensions of the country, large areas of degraded land and the still persisting cheap rural labour costs, create opportunities for the development of ambitious biomass electricity projects. Furthermore, climatic conditions seem to offer additional opportunities for solar and wind sources of energy, while remote natural gas supplies are still to be fully explored. Some of them, as we will analyse later, are not profitable on a private basis and the CDM credit can be very important in helping their implementation.

Forestry in Brazil is a much more promising area. Brazil has excellent opportunities in both carbon sequestration projects and the enhancement of sinks. The advancement of the agriculture frontier, based on forestland conversion financed by logging activities, has created large areas of degraded land where those activities are no longer privately profitable. These areas have been pointed out as a cheap land factor for forest plantations for pulp, timber and, as already said, biomass energy. The country's advanced technology in planted forests and climatic conditions makes those opportunities very attractive, although capital constraints have been a serious limitation to their expansion since they require long-term financing mechanisms.

The protection of native forest from deforestation can save much more carbon emission per area of land than any other carbon sequestration option. Sources of deforestation are not only due to regional development programmes, but also, if not mainly, to economic structural problems, such as land tenure and labour market imperfections. Nevertheless, deforestation control would only be effective with the closure of the agricultural frontier.

Profitability and Carbon Prices

Our economic analysis will estimate the current private profitability of each option and analyse how this profitability will be affected by carbon prices established through potential CDM markets. Carbon saved in each option is measured using its net present value. It is important to note that although some authors stand for the importance of discounting carbon through time,[4] here we are using a zero discount rate. This choice means that each option is generating the highest accumulated carbon savings and consequently that would result in the lowest possible carbon price in the market.

The reasoning is quite simple. We are analysing profitability of projects at the individual entrepreneur's level and this agent will not be interested in any social issues related to intergenerational welfare. Consequently, the issue of saving carbon today versus tomorrow only enters his decision if this affects his dynamic optimization path or if it is imposed exogenously by a protocol.

Carbon savings are measured as the emission reductions of the investment minus the emissions generated by a baseline activity that the investment is assumed to replace. Private profitability is measured for each option by the internal rate of return since it is the financial parameter guiding private investors. The present value of net revenue (ie, deducted of costs) divided by carbon saving indicates the minimum carbon price level that the option is willing to obtain at the CDM market. For negative net present values, the carbon price that will recover costs is calculated, corresponding to the carbon break-even price for the option at a certain rate of discount.

All financial estimates are based on a rate of discount of 12 per cent. Rates between 10 and 12 per cent rate can be seen as the common reference for opportunity cost of capital in Brazil and are assumed by international agencies. That reflects both a developing nation's risk premium of 2–3 per cent plus an overvaluation of the exchange rate of approximately 15–20 per cent. Following that, the efficient discount rate would be around 6–8 per cent, which is close to international standards. Here we assume that the CDM project's investors will require, in addition, a premium for CDM surveillance. This is the case because CDM projects will be certified and will be under external supervision, which increases risk and uncertainty about the expected revenue.

Land costs and oil prices are key parameters affecting the profitability of the options. In 1997, both land and oil prices were at the lowest level in the decade; since then oil prices have risen sharply, changing the assessments considerably. The results reported here are based on 1998 oil prices. Although this price level may not apply presently, the method remains valid when taken as a conservative estimate of the project profitability based on the fact that GHG emission-reduction projects are less economically attractive, the lower the oil price. To underscore the conservative character of the oil price assumptions, we assume that private investors will be risk averse and will tend to accept the current low oil price, and this, together with assumptions about relatively high past land prices, is used as an input to build up a worst case scenario.

Forestry Options in Brazil

The deforestation process in Brazil, as previously mentioned, has created a vast availability of degraded areas. These areas constitute a relatively cheap land stock that can be used for pulp, timber and energy biomass plantations. Nevertheless, as will be seen later, relative land prices in Brazil do not reflect this land stock for degraded land since the cost of clearing is part of the property titling process. Furthermore, although the country's advanced technology in planted forests, cheap inputs and good climatic conditions make those land use opportunities attractive, the development of these has been additionally limited by capital constraints and by the absence of long-term financing mechanisms.

On the other hand, the protection of native forest from deforestation can save much more carbon emission per area of land than any other carbon sequestration option. The curbing of the deforestation rate can move the country's pattern of GHG emissions away from the actual trend and save GHG emissions. Nevertheless, this is not an easy task.

Deforestation in Brazil is not only due to regional development programmes (as it was once heavily emphasised in the literature),[5] but also, if not mainly, to economic structural problems. Therefore, deforestation control is only effective with the closure of the agriculture frontier. Can a project based programme achieve this result?

Our analysis, surprisingly, indicates that a combination of sustainable logging, conservation and CDM credits can somehow be effective in this direction.

Timber exploitation in the Amazon takes advantage of legal land clearing which today allows 20 per cent of the property area to be cleared and which gives the right to deforestation.[6] Timber sales based on this licence provide an opportunity for raising up-front capital for full clearing afterwards and to bear later costs of securing property rights. Timber exploitation, in fact, in some areas acts as an inducing factor for land conversion.

Apart from the generally weak capacity of public agencies in a country where public deficit cuts are erratic and, sometimes, drastic, institutional performance in such large and remote areas is likely to be fragile and creates more room for illegal logging at the top of the clearing licence loophole.

A very minor proportion – around 10 per cent – of forestland in the Amazon region is suitable for crops and livestock. Cleared land eventually ends up being used for extensive cattle-raising in order to secure property rights. Once soil is degraded the movement to new areas of forestland continues. In a simple way, that has been the land-conversion pattern elsewhere, but in the case of the Amazon there is still time and sufficient opportunities to make better use of forestland.

National Forests (FLONA – Florestas Nacionais) in Brazil are conservation units for the purpose of sustainable logging. Of course, sustainability in logging activities is not possible without an environmental loss. Nevertheless, some field studies in the region have shown that logging using distinct practices can reduce these losses considerably.[7] The main problem is financial. Although 'sustainable' logging reduces production losses and increases tree growth, it results in higher costs than depleting logging as currently undertaken in the region, if land costs are taken into account. Sustainable practices require at least a 30-year rotation period, whereas unsustainable logging takes advantage of forestland clearing.

The FLONA concept is appealing. It can create, together with national parks, biological and extraction reserves, a quasi-sustainable land use in the Amazon. It can even stop the escalation of deforestation, making forestry a sound business. However, note that the synergy between forestland conversion to agriculture and timber production, which takes place in the region today, in fact is very cost-effective since extraction costs are already taken into account in the clearing process to open agricultural areas. On the other hand, sustainable logging in FLONAs has to be rotated, is capital intensive, has to be scientifically planned, and incurs recuperation and mitigation costs, and auditing and certification expenditures, given current labour laws and tax schemes. Sustainable logging based only on timber values may not be competitive, therefore, and its implementation may require additional sources of revenue.

Consequently, FLONA needs a by-product to cover the revenue difference in order to be competitive with illegal timber exploitation. The CDM may create an opportunity to cover that difference, as our analysis later on will try to show, although there are concerns about the efficiency of forestry projects designed as enhancement of sinks to achieve carbon savings. These arise mainly due to the large uncertainty about the effective long-term carbon savings (so-called leakage), since deforestation may take place somewhere else or forests may even catch fire due to natural reasons.

Next, the five main forestry response options in Brazil are described with an indication of their financial attractiveness. These options are: silviculture plantations for pulp, charcoal and sawlog, and sustainable native forest management for sawlog on a private and public land basis. Scale effects are only considered for land factors since it is the most relevant parameter in the forestry option. Learning costs are assumed to be small for plantations since they are already undertaken on a large scale in Brazil. Analysis of native forest management, however, is mostly based on field surveys since such activity is not undertaken in Brazil as yet. As a result, we introduce this analysis at project and simulated programme levels.

As shown in Table 8.1, all types of plantations have presented very high rates of return. The sawlog option is the most profitable with an internal rate of return (IRR) of 17.6 per cent, followed by pulp with 14.6 per cent, and charcoal with 13.3 per cent.

It is known that even when logging is undertaken without total land clearance, it harms the forest. Timber extraction using sustainable management practices aims at minimizing this ecological damage to forests and to reduce wood losses. On the other hand, it requires more land factors to each unit of wood produced in order to allow for rotation. Recent field survey studies on sustainable logging by Amaral et al (1998) and Almeida and Uhl (1995) admit that a 30-year harvest cycle is possible in dense tropical forests in Brazil to accomplish the same output per hectare as the one generated by current non-sustainable loggers.

254 Climate Change and Sustainable Development

Table 8.1 *Forestry response options in Brazil*

Project type	Planting for pulp in degraded areas	Planting for charcoal in degraded areas	Planting for sawlog in degraded areas	Private sustainable forest management for sawlog	Public concession forests for sawlog
Average carbon benefit tonnes C per ha[1]	24.1	180.0	43.3	18.0	18.0
Internal rate of return (%pa)[2]					
– excluding land costs	14.6	13.3	17.6	33.0	–
– including land costs	11.1	10.1	13.3	0.5	1.3
Net present value per tonne of C benefit (US$/tonne C)[3]	1.4	0.7	–9.5	9.0	1.8

Source: Based on Serôa da Motta, Ferraz and Young (2000)
1 The carbon benefits from plantations are measured as the difference between the carbon stock of plantations and the carbon stock of degraded land. For sustainable timber management, it is measured as the difference between the carbon stock of sustainable logging and non-sustainable logging plus wood products, soil carbon and fossil fuel substitution (in the case of charcoal), according to estimates from Fearnside (1995)
2 Estimates for plantations without land costs are taken from Fearnside (1995) and sustainable management values in column 4 for the case without land are taken from Almeida and Uhl (1995). Land cost inclusion (US$200/ha) is elaborated by the authors considering a perpetuity. All values are in US$ 1992
3 Measured as the net present value divided by the carbon benefit for each option. Values give the carbon break-even price to make the internal rate of return equal to 12 per cent pa. Positive values mean that investments are already profitable even if no account is taken of the carbon benefits

However, given the open access that characterizes the logging pattern in the Amazon, loggers do not invest in land since they rely on land clearance resulting from frontier expansion. Consequently, our comparison is based on sustainable logging with land investment and a 30-year rotation versus unsustainable logging without land purchase.

The unsustainable logging process will harvest 38m³/ha and then move forward over the frontier.[8] On the other hand, sustainable logging can generate a yield of 1.27m³/ha. Therefore, in order to achieve the same output per hectare as open-access activity, sustainable loggers will have to buy 30ha of land to comply with their management plans.

Here we analyse the option of sustainable forest management considering a project-level approach, where land is acquired at the market by private loggers. Later we analyse other possibilities related to public land on a concession basis for large tracts of land for timber production in the Amazon.

Carbon benefits of US$18 per tonne of carbon per ha for sustainable management were measured as being the difference between carbon stock from sustainable logging and carbon stock from non-sustainable logging, according to estimates presented by Fearnside (1995).

Based on these estimates, private sustainable timber management, including land costs, is the option that will require the highest carbon price – namely, around US$9 to make it viable, as shown in the table. The option with the lowest implicit carbon price is silvicultural plantation for sawlog that even has a negative price and thereby is profitable without taking the value of carbon savings into consideration.

The National Forest Programme designed by the current government has proposed concessions for 40–60 million ha to be exploited through sustainable management practices that, based on a rotation period of 30 years, generates an output level of 50–70 million m³ of roundwood.

Therefore, one option to be considered is a project that makes use of public land availability in the Amazon based on a concession system. According to Funatura (1992), almost 35 per cent of forestland in the Amazon is publicly owned. The same study has proposed an ambitious plan to create conservation units which, added to the existing ones, would cover 30 per cent of the Amazon Forest, including 60 million ha of national and state forests under concession schemes.

Since the specific plan for concession-covered forests is based mostly on public land and aims at an area of 60 million ha, the estimated average cost, including operational and maintenance expenditures of the units, was estimated at US$69.50 per ha – ie, much lower than the price of US$200 used for private enterprises. Taking into account the net price and costs for sustainable management presented above, the concession scheme, at this land cost, would generate a rate of return of 1.3 per cent and a carbon break-even price of US$1.80, as shown in Table 8.1.[9]

As can be seen, CDM carbon revenues from concession forests can be very important since these net revenues are expected to provide payments for forest capital depreciation – setting aside proceeds to recover forest-user costs.

Energy Options in Brazil

Although hydroelectricity still accounts for almost the total supply of electricity in Brazil, the hydropower stock in Brazil already has been mostly taken up. The very few sources still available will face high production and distribution costs. For example, most of them are in the Amazon which will represent high production costs in terms of reservoir building and ecological damage plus the cost of taking energy from these sources to the big cities in the south, more than 3000km distant.

The CDM provides unique opportunities to reduce, at least partially, the damaging consequences of large-scale adoption of fossil fuel thermoelectricity in Brazil through the creation of economic incentives for the adoption of alternative renewable energy options. The objective of this section is to present some of these options, providing preliminary estimates of the best energy options in terms of their contribution to combat climate change.

Energy options have been analysed here based on the current price paid per MWh electricity to independent producers and the current gasoline price in the case of ethanol production. For plantation options, assumptions about carbon benefits and wood costs were the same as applied for forestry options.

Due to data differences, energy options were assessed based on 1998 US$ prices, whereas forestry options were based on 1992 US$ prices. Adjustments to equalize the currency standards were not applied since the variation of the real exchange rate in the period was less than 5 per cent.

The options analysed were: ethanol fuel with cogeneration of electricity; cogeneration of electricity from industrial chemical plants; fuelwood gasification from pulp residues; and wind energy.[10]

One strategy increasingly pursued by producers in order to lower ethanol production costs is the use of sugarcane residues to produce steam and/or electricity. Conventional low-pressure steam boilers are very inefficient, but there are important innovations in this field. The most promising is the gasification of biomass residues, allowing the use of high-performance gas turbines. Even though this is a new technology, there are already investment decisions being taken on its use. Indeed, if the raw material of this process is considered costless (being a necessary output of sugar and ethanol production), cogeneration can be a profitable activity. There are projections for the state of São Paulo (higher levels of output and productivity in Brazil) that cogeneration capacity will increase by 180MW by 1999 (IE, 1998). Therefore, cogeneration, at the same time, will replace natural gas or fuel oil consumption (that are expected to accommodate, at least partially, the growing energy demand in Brazil) and will reduce the aggregate costs of ethanol production.

The results of a negative rate of return and a carbon price of US$19.70 observable in Table 8.2 may indicate that this option will not be so financially attractive, even if one considers cogeneration and large carbon savings.

In our exercise, we considered that 1 tonne of bagasse requires 4 tonnes of sugarcane, and generates 290kWh at an efficiency of 40 per cent.[11] Capital costs were considered at US$1400/kW (based on the cost of the same technology for fuelwood gasification).[12] The occupation ratio used was 80 per cent, being used 190 days per year (the supply of bagasse is seasonal, concentrated in the harvest months). The energy produced was considered to replace either natural gas or fuel oil consumption, avoiding the emission of 106kg.C/MWh and 180 kg.C/MWh respectively. Finally, it was considered a carbon net emission of 14 tonnes C/ha in 50 years, assuming that land has to be converted (from pasture) to allow for the expansion of sugarcane plantations.

There are other sectors in which cogeneration has been growing steadily. The most important are the chemical and petrochemicals industries, pulp and paper, and metallurgy. The existing capacity (officially recorded within the electricity companies) is around 1100MW, but there is potential for much more: the potential energy surplus can exceed more than half the current electricity generation in Brazil.

Table 8.2 reports the cogeneration potential of a refinery. The large scale of these plants in the chemical industry is an advantage for cogeneration projects that are very capital intensive, and ongoing investments in these plants have been planned through the whole sector. CDM funding may enhance the range of projects that are financially feasible, reducing the need of further expansion of thermoelectricity (thus saving fossil fuel emissions).

The results reported in Table 8.2 confirm previous studies that indicate cogeneration as a priority option for market-oriented, carbon-saving investments (considering that energy surpluses replace the expansion of fossil fuel thermo-electricity). Among the options analysed, it presents the highest internal rate of return over 20 per cent. Since cogeneration improves the use of inputs, which would be consumed anyway, it does not generate further externalities. The main environmental impact is limited to carbon savings from avoided emissions in thermoelectricity, but its high profitability would allow it to capture any carbon price. There are no significant indirect social impacts (job creation is very low).

As already mentioned, there are important technological improvements in biomass energy systems. Research on areas such as biomass gasification from fuelwood, agricultural residues and urban waste indicate that commercial units would perform reasonably in economic terms. Gasification improves the system efficiency and in many cases is more than doubling the output of energy per unit of biomass (IE, 1998). The case of a wood biomass commercial unit has been considered here, similar to the experience of the WBP/SIGAME project in Bahia (funded by the GEF). This project assumed that half of the total raw materials comes from residues from the pulp industry.

The result shown in Table 8.2 indicates a moderate positive rate of return of 8.5 per cent and a carbon price of only US$2.40 per tonne for the gasification

Table 8.2 *Energy response options in Brazil*

Project type	Fuel ethanol plus electricity cogeneration from residues	Fuelwood gasification with pulp residues	Electricity cogeneration from chemical plants	Wind energy
Average carbon benefit:[1]				
tonnes of C per ha	203.200	102.300	–	–
tonnes of C per MWh	0.056	0.197	0.18	0.18
Internal rate of return (% pa)[2]	negative	8.500	21.90	7.30
Net present value ($US/tonne C)[3]	19.700	2.400	−17.70	14.60

Source: Based on Serôa da Motta, Ferraz and Young (2000)
1 The carbon benefits of the ethanol, the cogeneration and the gasification options are the difference between the carbon stock of plantations and the carbon stock of degraded land as measured by Fearnside (1995) plus fuel oil substitution from COPPE (1998). Carbon benefits from electricity from oil refinery and wind sources are measured from fuel oil substitution from COPPE (1998)
2 Rate of return estimates for ethanol and cogeneration and gasification consider land costs of US$200.00. All values are in 1998 US$
3 Measured as the net present value divided by the carbon benefit for each option. The figures imply a carbon break-even price that generates a rate of return equal to 12 per cent per annum. Positive values mean investments already profitable at this rate. All MWh revenues were assumed to be the current price paid in the Brazilian electric sector of US$35 per MWh

project. This suggests that private entrepreneurs may invest in this option if the plants are located in remote areas (where energy costs are higher) or if supply shortcuts are foreseen (thus avoiding production interruption due to lack of energy). Larger scale adoption of the gasification option, however, may require some external funding. Potential environmental impacts may happen if proper measures are not considered to control water and air emissions. The exercise assumes that the fuelwood is obtained from the conversion of degraded land to managed forests. The carbon saving, therefore, exceeds the carbon saved exclusively from avoided emissions from thermoelectricity expansion.

Among the 'clean' energy technologies, the wind energy option appears as one of the most promising in Brazil. It presents lower costs than solar units (the 'rival' alternative option which is particularly attractive in remote areas), even though wind energy requires a relatively high degree of technical expertise. The north-east region presents the better perspectives and one commercial unit in the state of Ceará commenced normal operation in December 1998, selling electricity at competitive prices (IE, 1998). Most of the data used here is based on this experience, and it is concluded that wind energy is economically attractive, even though extra funding, such as supplied by the CDM, would considerably increase the range of viable units. Due to the relatively low carbon-saving capacity of wind turbines compared with the other options that also imply increased carbon sequestration by decreased afforestation or similar biomass practices, wind turbines require a high carbon price of US$14.60. The positive secondary effects of wind turbines are the avoided emissions of both local and global pollutants resulting from fossil fuel thermoelectricity. Other environmental impacts and job creation effects are negligible. The most important impact on social development is the possibility of generating electricity locally in remote areas where it would not come otherwise because of the high transmission costs.

Identifying the Most Attractive Options

As noted earlier, the profitability of energy options is very sensitive to the current oil prices. The higher the oil prices the lower will be carbon prices required for making the options economically attractive, although the relative strengths of options may not change.

Other important factors include future fossil fuel prices as well as land prices. Consider, for example, the case of charcoal. Currently, charcoal production is only economically viable given a very low carbon price based on the current charcoal market prices. However, charcoal prices may come down fast since these will be highly influenced by coal prices, which currently are declining. Such scenarios have to be analysed before a final conclusion on any one option's profitability can be made. This is, however, beyond the scope of the current study.

On this basis the preliminary conclusion is that the worst case scenario, assuming low oil prices and high land costs, suggests four options as the most attractive CDM options:

1   Electricity cogeneration from chemical plants
2   Silvicultural plantations for pulp, sawlog and charcoal
3   Gas production from firewood and pulp residues
4   Sustainable logging in native forests

In terms of generating carbon savings for CDM transactions, sustainable logging activities in national forests in the Amazon, in the scenario of replacing all open-access logging in the region, would generate 18 tonnes of C per ha and that could take place in an area of 60 million ha.

Considering the existing industrial capacity, electricity cogeneration from chemical plants could generate 31,682GWh and save 0.180 tonnes of C per MWh. Gasification from firewood could count on 5 per cent of pulp residues, which today would be in the order of a magnitude of 1.7 million m³ . This quantity could generate almost 1.8GWh of electricity, saving 0.197 tonnes of C, including carbon stocks from afforestation. We assume a project lifetime for both options of 50 years.

Silvicultural plantations, considering the current non-utilized land area in the states at the border of the Amazon, could double its current annual production level of 106.6 million m³ and generate an average 88.2 C per ha over the total project lifetime.[13]

However, since cogeneration and silviculture plantations require by far the lowest carbon prices, they would be the first to take their shares, which, combined, account for more than a third of total expected carbon.

If the carbon price drops to very low levels, the CDM market will reject sustainable logging and biomass gas in favour of industrial cogeneration and plantations. Would that be of any significance for Brazil's welfare? The answer will depend on secondary benefits arising from carbon-saving projects.

Secondary Benefits

The analysis of secondary benefits will be undertaken for the options discussed above. These will be compared with each other and also, when applicable, with their probable alternative – for example, plantation against extensive pasture, sustainable against non-sustainable logging, wind energy against thermoelectricity.

It is certainly complex and controversial to evaluate secondary benefits, even when one assigns monetary values in order to make them comparable. The indicators in Tables 8.3 and 8.4 are only qualitative indicators that give a sense of the direction of these secondary benefits and their relative magnitudes. Each type of option is classified in a two-entrance matrix where the first entry represents the direction of the secondary benefit (positive or negative) and the second its relative magnitude (low, medium, high).

It is important to note that this classification does not attempt to cover the numerous aspects arising from each options, but only reckoning on some of them which are more relevant and directly identified. Furthermore, to offer a sense of aggregated evaluation, we have highlighted the highest positive indicator in each benefit for each relevant option.

Environmental Quality Benefits

The following quality benefits are considered.

Water resources availability measures the effect of each option on relative water scarcity.

For forestry projects, apart from the semi-arid region at the north-east of Brazil, the country's water availability is great with high scarcity levels only present in high-density urban and rural regions. Irrigation is a high national priority, but none of the options analysed above will increase water competition from irrigation. The debate on water impacts from plantations has pointed out that plantation activity reduces ground water availability when compared to pasture, although no definitive evidence has been produced to confirm that. According to Chomitz and Kumari (1996), it is also poorly understood how the deforestation of native forest affects hydrological conditions. Based on that, one cannot assure that sustainable timber management will grant significant water availability benefit. Nevertheless, some weak evidence suggests that plantation options, as analysed here, can generate potentially negative impacts on water availability, whereas sustainable logging could do the opposite, avoiding water depletion, particularly if undertaken at such large areas as planned with concession forests.

For energy projects, sugarcane plantations are other relevant sources of water demand and could create additional pressures on water scarcity in some regions. Additionally, wood plantations for energy have similar impacts to those described for timber. Cogeneration (refinery) and wind energy have no direct significant impacts on water scarcity.

Water resources quality measures the effect of each option on water pollution or the assimilative capacity of water.

The use of chemical input in plantations has clear negative impacts. Brazilian rivers, as elsewhere in the world, are already showing high indicators of phosphorus and nitrogen in some rural areas. If plantations are due to be implemented near these areas, the environmental outcome is certainly negative. On the other hand, sustainable logging is free of such a negative impact on water quality.

On the energy side, ethanol production, if not properly managed, may result in serious water pollution problems given the generation of residues that are rich in organic matter. Modernization within the sector, however, has resulted in making good use of these residues as fertilizers in the sugarcane plantations. Other problems are organic residues from the washing of sugarcane before processing (not only increasing the water-emission problems, but also reducing the productivity – an estimated 3 per cent of sugar is washed away with this practice) and chemical emissions from their intensive use in plantations. All these problems are avoidable with more efficient production techniques.

Wood plantations for energy have similar impacts to those described above for forestry. On the other hand, cogeneration (refinery) and wind energy have no significant water resource impacts, but they may lead to the avoidance of problems from conventional thermoelectric generation.

Urban air pollution measures the effect of each option on air pollution or air assimilative capacity.

Since it was assumed that sustainable management will not curb forest burning, it is not possible to consider air pollution benefits to the cities around land-clearing areas. Urban pollution in big cities in Brazil is high, but it has been caused mainly by mobile sources. Industrial sources have been placed away from populated areas, apart from very specific cases.

Table 8.3 *Qualitative indicators of secondary benefits for forestry options (sign and magnitude)*

Secondary benefits	Pulp plantation in degraded area	Charcoal plantation in degraded area	Sawlog plantation in degraded area	Private sustainable native forest management for sawlog	Public concession forests for sawlog
Environmental impacts					
Effects on water resources availability	negative low	negative low	negative low	positive medium	positive high
Effects on water resources quality	negative low	negative low	negative low	neutral	neutral
Effects on urban air pollution	negative low	positive low	not relevant	not relevant	not relevant
Effects on soil erosion	negative uncertain	negative uncertain	negative uncertain	positive high	positive very high
Effects on biodiversity protection	positive low	positive low	positive low	positive high	positive very high
Development impacts					
Effects on aggregate demand	positive very high	positive very high	positive very high	positive high	positive high
Effects on trade balance	negative low	neutral	negative low	positive medium	positive medium
Effects on regional economy	positive low	positive low	positive medium	positive very high	positive high
Opportunity cost of output forgone	negative low	positive low	negative low	negative low	negative medium
Equity impacts					
Effects on income distribution based on the project's unskilled labour participation	positive low	positive low	positive low	positive high	positive medium
Effects on the consumption of the project's output by income class	neutral	neutral	neutral	positive low	positive medium
Effects on the distribution of environmental benefits by income classes	neutral	positive low	neutral	positive medium	positive high

Source: Based on Serôa da Motta, Ferraz and Young (2000)

Table 8.4 *Qualitative indicators of secondary benefits for energy options (sign and magnitude)*

Secondary benefits	Ethanol (with bagasse cogeneration)	Cogeneration from refineries	Biomass thermoelectricity (gasification of wood)	Wind energy
Environmental impacts				
Effects on water resources availability	negative low	not relevant	negative low	not relevant
Effects on water resources quality	negative medium	not relevant	negative low	not relevant
Effects on urban air pollution	positive low	positive medium	negative uncertain	positive high
Effects on soil erosion	negative medium	not relevant	negative uncertain	not relevant
Effects on biodiversity protection	uncertain	not relevant	positive low	not relevant
Development impacts				
Effects on aggregate demand	positive high	positive low	positive medium	positive low
Effects on trade balance	positive high	positive high	positive high	positive low
Effects on regional economy	positive very high	positive low	positive high	positive medium
Opportunity cost of the output forgone	positive low	neutral	positive low	neutral
Equity impacts				
Effects on income distribution based on the project's unskilled labour participation	positive high	neutral	positive low	neutral
Effects on the consumption of the project's output by income class	negative medium	neutral	positive low	neutral
Effects on the distribution of environmental benefits by income classes	positive high	positive medium	positive low	positive medium

Source: Based on Serôa da Motta, Ferraz and Young (2000)

The industrial use of charcoal can generate emissions less harmful to human health than coal and other fossil fuel, whereas emissions at the production centre can be a serious problem if carried on without proper technological devices. Since charcoal substitution will positively affect highly populated and polluted industrialized areas and charcoal production will pollute small cities, it is plausible to assume that the net benefit on air pollution from charcoal production is positive.

Pulp production is known to cause distress to neighbouring areas since it produces significant odour effects from air emissions.

For energy projects, the pre-harvest burning practice, essential in sugarcane manual harvesting, results in serious air pollution problems if plantations are near urban centres.[14] However, if the leaves and other parts of the sugarcane traditionally burnt are used for cogeneration instead, there may be an improvement in electricity performance. The problem is the requirement of mechanical processes for doing so, leading to considerable loss of jobs.

Biomass burning may also result in air pollution problems. Process optimization and end-of-pipe practices are required in bagasse and wood projects, even though gasification reduces the level of air emissions. The initial advantage of ethanol-fuelled vehicles being less polluting than gasoline ones in terms of local pollution parameters (non-CO_2 emissions) cannot be regarded any more because of the improvement of engine efficiency and catalysts.

Cogeneration (refinery) and wind energy have no significant impacts on their own, but they may lead to the avoidance of problems from conventional thermoelectrical generation, which are potentially serious sources of urban air emission problems.

Soil erosion control measures the effect of each option on soil erosion or land protection capabilities.

Again, plantations have a clear negative impact on soil erosion since they are a mono-culture land practice and they also increase soil acidification. However, compared with pasture, soil erosion from plantations is lower – a benefit. Sustainable logging, on the other hand, certainly has a positive soil impact since it reduces vegetation losses compared to open-access logging, particularly at the scale of concession forests.

On energy-related projects, soil compacting is a consequence of persistent land use in sugarcane plantations. Additionally, wood plantations for energy present similar problems as the ones presented above for forestry.

Biodiversity protection measures the effects of each option on biodiversity protection.

As discussed in the previous section, sustainable logging is by far the option with the highest benefit on biodiversity protection. As pointed out by Amaral et al (1998), even considering the ecological losses, they are minimal compared to the current non-sustainable practices. In addition to that, sustainable logging may reduce deforestation and open new areas for scientific research on biological genetics. On the other hand, plantations will gain too little biodiversity values when compared to pasture land.

The effects on energy projects are only relevant for ethanol. If sugarcane expansion occupies previously deforested areas, the impacts on biodiversity are neutral, but if there are direct or indirect incentives for deforestation, the impact would be potentially higher.

Development Benefits

Effects on aggregate demand measure the back and forward linkages in the economy resulting from project implementation.

All forestry options will generate a similar demand for machinery. However, plantation will increase demand for seeds, chemical input and land at a much higher rate than sustainable logging and its alternative land uses. In addition, genetic research is more in demand from plantation than from native forest management.

Forward linkages are very similar if one considers that wood products can be adapted to either planted or tropical timber, although they are not really perfect substitutes.

On the other hand, ethanol production has important multiplier effects, given its relatively higher employment capacity and its importance to local economies. Cogeneration and wind energy tend to present few (if any) linkages.

Effects on the balance of trade measures the import and export effects from project implementation.

The growing negative trade balance in Brazil has been a serious macro-economic constraint. Assuming that an important part of the additional timber produced from sustainable logging is going to be exported, there will be a secondary benefit in terms of improving the trade balance. The higher input demand of plantations, however, can impose a higher burden on trade deficit than native forest management, if it is applied at large scale by increasing imports. Charcoal would be the exception since additionally it avoids fossil fuel imports.

From the energy point of view, most of the practices described here are import saving (as far as they replace imports of fossil fuels). Net results for wind energy are less significant because it is possible that machinery and equipment have to be imported.

Effects on regional economies measures the proportion of income generated from the project implementation, which stays at the regional level.

Plantations and sustainable management are equally important to regional economies. However, again, the high input demand from plantations will reduce the value added captured in the region. Moreover, pulp and charcoal industrial uses may occur elsewhere, whereas sawlog tends to be processed in the region. If international capital plays an important role in concession forests, it may also reduce value added captured in the region.

On the energy side, the impacts of ethanol production are very important to local economies, by job creation and by infrastructure building. The main advantage of other sources, particularly biomass and wind energy, is to allow rural electrification (an important instrument for development) in remote areas, where electricity would not be economically feasible otherwise, given higher transmission/transportation costs.

Opportunity costs of the output are the opportunity costs of activities which are forgone due to the project implementation.

Extensive pastures which are assumed to be displaced by plantation generate a very low value added when compared to non-sustainable logging. Charcoal substitution may reduce fossil fuel production elsewhere. Therefore, the output forgone per hectare from sustainable logging can be much higher than from plantations, particularly at the scale of concession forests. Nevertheless, long-term sustainable logging, since it can generate a perpetual timber income, is a more interesting use of land than the current logging practices.

Sugarcane and biomass electricity have similar effects as charcoal, whereas other options tend to be neutral.

Equity Benefits

Effects on income distribution based on the project's unskilled labour participation measures the effects of each option on the relative demand for unskilled labour as a way to address the possible effect on income inequality.

All forestry options will present similar labour intensity. It is expected, however, that plantations will require relatively more qualified workers than sustainable logging since it relies on more sophisticated techniques. Sustainable logging, on the other hand, can rely mostly on indigenous people and communities in field activities, despite the fact that large concession forests may be less prone to that than private sustainable logging.

Although sugarcane plantations are intensive in unskilled labour, mechanization may improve productivity, at the cost of forgone jobs – this is a dilemma yet to be solved for the sector. Cogeneration and wind energy do not present significant employment and equity effects.

Effects on the consumption of a project's output by income class appropriation by income classes of the marketed output generated by the project.

Sustainable native forest management has been shown in this analysis to compete with open access activity at low wood prices. Nevertheless, part of this low-price timber will be exported. Plantations are just an expansion of the current production pattern. The effects will probably be neutral since all consumers will benefit symmetrically from an increase in the supply.

Energy generated by ethanol will be used mainly by those with higher incomes, so the benefits from consumption will be distributed regressively.

On the other hand, energy in general will be consumed more symmetrically across society, so the impact of the project on final product consumption is expected to be neutral across income classes (except, of course, for sections of the population who do not have access to electricity).

Effects on the distribution of environmental benefits by income classes measures the appropriation by income classes of each project's secondary environmental benefits.

The large environmental benefits from protecting the biodiversity of sustainable logging will be captured equally by all income classes. Charcoal plantation has potential positive environmental impacts, whereas all types of plantations will affect water degradation negatively. As with any pollution problem, it harms the poor most since they cannot afford defensive expenditures. However, water benefits from sustainable management may relatively favour the low-income classes.[15]

From the energy projects perspective, pollution generated in sugarcane plantations will also relatively harm the low-income classes, but it can improve air quality in big urban areas. Biomass electricity replaces thermoelectric projects and thereby reduces air pollution concentration, which affects all income groups.

Final Comments and Recommendations for Brazil

As pointed out in this analysis, private investors seeking CDM rents will be more willing to undertake plantations than native forest management, since the former offer higher profitability against lower learning costs, reasonable scale effects and, above all, lower carbon break-even prices. CDM buyers would also opt for plantation due to low leakage rates.

Industrial cogeneration of electricity is by far the option with the highest private return. Biomass electricity and wind energy options are equally profitable compared with plantations. Ethanol production, however, is not privately profitable, and it would only be a CDM option if it is supported by a government intervention.

When secondary benefits, however, are taken into account a different picture may arise from the current analysis.

Looking at Tables 8.3 and 8.4 (see p261–262), it can be observed that in terms of environmental benefits, native forest management options – particularly concession forests – offer a great deal of secondary benefits with great relevance for biodiversity protection. Biomass electricity as a substitute for charcoal consumption can also assure air pollution benefits.

For development impacts, plantations are more important for the activity level of the economy as a whole, but less for the regional economy, although they can affect the trade balance deficit negatively. In terms of regional benefits, private sustainable logging in native forests is more relevant. Ethanol and biomass electricity, on the other hand, capture most of all development gains.

Equity issues are in favour of native forest management, when they affect low-income classes through the project's output, costs and ecological benefits, although they generate more negative impacts from displacement activities than plantations.

Ethanol and biomass electricity combine development gains with equity ones. In the case of ecological impacts, wind energy offers more air pollution benefits, while biomass electricity affords more protection of biodiversity.

This partial qualitative analysis is summarized in Table 8.5, which presents generalized indicators of secondary benefits for the previously analysed options. As can be seen, private profitability, by itself, has no definitive linkage with secondary benefits – ie, market forces alone will not be able to select CDM options, which, at the same time, have high private cost-effective and high positive linkages to ecological and social benefits.

If the CDM market offers a carbon price that makes several of these options attractive, including the ones with high secondary benefits, there is no reason to

Table 8.5 *Generalized indicators of benefits from response options in Brazil*

	Industrial plantations and biomass electricity	Sustainable forest management	Ethanol with electricity cogeneration	Industrial cogeneration of electricity	Wind energy
Private returns	medium	low	very low	high	medium
Environmental benefits	low	high	medium	low	high
Development benefits	high	medium	high	high	low
Equity benefits	low	high	high	low	low

Source: Based on Serôa da Motta, Ferraz and Young (2000)

be concerned about the country's welfare maximization by undertaking these projects, since all the benefits will be captured.

As can be seen, the conflict between private profitability and welfare maximization arising from CDM opportunities is clear and there is no simple approach which can solve it. For the time being, the CDM debate has to be enhanced to grasp the benefits of mixing carbon savings with other ecological and social benefits in order to maximize both the global carbon-saving benefits and the welfare of the project host country. Presently, CDM design and procedures are far from being concluded and one can expect much controversy in their resolution. However, as this analysis has shown, the reconciliation of carbon-saving cost-effectiveness with countries' well-being can be one of the key factors to reduce controversies and make the most of CDM opportunities, slowing down climate change and enhancing the development process.

CASE STUDY FOR INDIA

Introduction

The current gross emissions of GHGs in India, on a per capita basis, are only one-sixth of the world average (ADB, 1994). However, despite this, India is one of the top ten contributors in the world to GHG emissions. The sector-wise inventory of GHGs for 1990 is presented in Table 8.6.

India, as a developing country, does not face any targets for reduction of GHGs under the Kyoto Protocol, but is able to avail of the CDM mechanism. It is reasonable to expect that a major share of the CDM activity in India would take place around GHG emission reduction in the energy sector, which accounts for more than 85 per cent of the CO_2 emissions in the country.

It is in this context that the present chapter analyses the various climate change mitigation options in India, keeping in mind its developmental priorities. To assess the overlap of Indian sustainable development priorities and the

268 Climate Change and Sustainable Development

Table 8.6 India's national GHG inventory for 1990 (Gg)

GHG sources and sinks	CO_2 emissions	CO_2 removals	CH_4[b]	N_2O	NO_x	CO	CO_2[c] equivalents
1. Energy							
A Fuel combustion	508,600						
• Energy and transformation industries					2684[d]	3493[d]	508,600
• Biomass burning	300,460[a]		1579	11	400	11,492	36,569
B Fugitive emissions from fuels							
• Solid fuels			330				6930
• Oil and natural gas			626				13,146
Total from energy sector (A+B)	508,600		2535	11	3084	14,965	565,245
2 Industrial processes	24,200			1			24,510
3 Solvent and other products							
4 Agriculture			12,654	243	109	3038	341,064
5 Land use change and forestry (LUCF)	52,385	–50,900					1485
6 Waste			3288				69,048
Total national emissions and removals	585,185	–50,900	18,477	255	3193	18,003	1,001,352

a CO_2 emissions from biomass burning are not included in the national totals
b CH_4 emissions according to the IPCC 1996 methodology
c CO_2 equivalents are based on global warming potentials (GWPs) of 21 for CH_4 and 310 for N_2O. NO_x and CO are not included, since GWPs have not been developed for these gases. Bunker fuel emissions are not included in the national total
d NO_x and CO emissions are computed for the transport sector
Source: ADB (1998)

possible mitigation options, we review first the development priorities of India. The succeeding section then puts into perspective the possible mitigation options by highlighting the emissions in the Indian economy and possible future growth paths. This is followed by a brief discussion on mitigation options in India. The final section analyses some of these mitigation options and undertakes a prioritization exercise to rank the options based on a sustainable development perspective, which is based on the inputs of a wide range of stakeholders of the Indian economy. A multicriteria analysis (MCA) technique is used to undertake the analysis. This section is based on the work carried out under the ALGAS[16] project (TERI, 1998d; ADB, 1994).

Development Priorities in India

Despite considerable development in the Indian economy since independence in 1947, the basic necessities of a large section of the population are unmet. A slow rate of growth and a high rate of population growth over the first four decades since independence have resulted in slow improvement in living standards in the economy. A sizeable population lives in poverty and is deprived of access to basic amenities, such as sanitary facilities, drinking water and health facilities. The Ninth Five Year Plan[17] (1997–2002) highlights the development priorities of the country as:

• Accelerating economic growth while maintaining stable prices;
• Promoting agriculture and rural development;
• Improving the supply of drinking water and primary healthcare;
• Containing the population growth rate;
• Ensuring environmental sustainability of the development process;
• Empowering women and socially disadvantaged groups;
• Promoting and developing people's participatory institutions; and
• Strengthening efforts to build self-reliance.

A new economic perspective was launched in 1990 to increase the economic growth rates, stressing an open economy and a greater role of market forces in economic decision-making. These policies are aimed at increasing the rate of investment in the economy and moving the economy towards an efficient use of capital. The economic expansion is expected to increase the labour absorption and thus reduce unemployment, which is a major socioeconomic concern. The Ninth Five Year Plan (NFYP) states that:

> *a primary objective of state policy should be to generate greater productive work opportunities in the growth process itself by concentrating on sectors, sub-sectors and technologies which are more labour intensive, in regions characterised by higher rates of unemployment and underemployment.*

The agricultural sector still commands substantial attention in development priorities. The fact that around 60–70 per cent of the Indian population resides in rural areas and is dependent on agricultural and allied activities for income generation implies a greater need for ensuring rural and agriculture development to improve the standard of living of the total population. The NFYP states that:

> *Despite progress in the other areas over the past decades, we are woefully lacking in providing basic services such as health care, education, safe drinking water etc. to the majority of our population especially in rural areas.*

The stagnation in agricultural growth and the lack of development of the rural infrastructure to support rural industries has affected the standard of living in rural areas. This has also increased the pressure on urban areas through rural-urban migration. Along with the stress on augmenting the infrastructure to increase agricultural production and productivity, the stress is on creating a rural industrial base. An important priority of development is providing adequate and clean energy to rural areas.

Another major issue that has slowly gained recognition as an integral part of development is the environment. The process of development since independence has led to a significant degradation of the natural and environmental resources (TERI, 1998c). The *Asia-Pacific Environment Outlook* identifies water resources, industrial pollution, urban congestion, land and soil resources and deforestation as the major environmental issues for India (UNEP, 1997).

The major sources of air pollution in the country are industries (toxic gases), thermal power plants (fly ash and sulphur dioxide) and motor vehicles (carbon monoxide, lead and particulate matter) (TERI, 1998b). Solid waste includes heterogeneous urban household waste together with more homogeneous accumulations of industrial, agricultural and mining wastes. It is roughly estimated that Indian cities and towns generate about 4000 tonnes of municipal wastes every day, creating pressure for land space and generating air pollutants and other problems with their disposal. Nearly 90 per cent of waste was land-filled in 1991 (EPTRI, 1995). Overall, it is estimated that the industrial sector generates about 100 million tonnes of non-hazardous solid wastes – with coal ash from thermal power stations accounting for more than 70 million tonnes – and 2 million tonnes of hazardous waste a year (TERI, 1998b). Municipal sources contribute three-quarters of wastewater measured by volume; industrial waste makes up more than half of the total pollution load. Average figures for Class II towns indicate that about 90 per cent of the water supply is polluted and only 1.6 per cent of the polluted waste water gets treated (CPCB, 1990). Soil degradation and the loss of forest cover is another major environmental concern. The degradation of agricultural soils poses a major challenge for maintaining agricultural productivity. The major sources of soil degradation are soil erosion, due to both water and wind, soil salinity, the depletion of nutrients and chemical contamination. The loss of forest ecosystems and their degradation due to increased pressure from fuelwood, fodder and timber demand have been significant since independence (Ravindranath and Somashekhar, 1995). Demand for timber is predicted to increase from 30 million tonnes per year in 2000 to 43 million tonnes in 2020; firewood from 185 million tonnes to 272 million tonnes (Ravindranath and Somashekhar, 1995). Meeting these demands poses challenges for the Indian forestry sector, quite apart from the carbon implications, and, moreover, this pressure will only increase in the future.

One of the major sources of pollution is energy use and energy conversion. Energy is also an important driver of economic growth. The principle objectives outlined in the NFYP for the energy sector are to meet the energy needs of the economy through the efficient and sustainable use of resources. Between 1984 and 1994, final commercial energy consumption grew at a rate of about 5.6 per cent per annum and now stands at 450MTOE (million tonnes of oil equivalent) (IEA, 1998). India's energy GDP elasticity[18] is fairly high at 0.9–1.0, although it does exhibit a slight downward trend in recent times. While industry is presently the biggest user of commercial energy, an important source of future demand will come from the residential sector. Another major unmet demand source is in the rural areas – only a third of rural households are presently electrified. Rural electrification is high on the priority in the electricity sector agenda.

A major source of commercial power is electricity. India has an installed capacity of about 93GW in March 1999. Gross power generation from utility capacity has increased from 5 billion kWh in 1950–1951 to 448 billion kWh in 1998–99 – an average growth rate of about 10 per cent per annum. Even at this rate, capacity expansion has been unable to keep pace with demand due to serious financial constraints. Shortages remain fairly constant, with an energy shortage of 6.2 per cent and a peak shortage of 12.5 per cent in 1999–2000 (TERI, 2000). New capacity of 12GW will be required every year under the next three Five Year Plans and power demand by the year 2020 is estimated to be 385GW, requiring about US$380 billion to be invested in the power sector alone (TERI, 1998b).

The energy sector was opened to private investment post-1990 to meet the increased resources requirement for creating adequate energy infrastructure. Another major initiative has been on the efficient use of energy resources and demand-side management. The Plan Document outlines the need for developing institutional mechanisms for energy conservation, demand management, research and development efforts and regulatory mechanisms towards achieving these goals. In line with past reforms of energy pricing, tariffs will be rationalized and the use of 'time-of-day' metering and peak-period pricing will be explored with a view to eventual implementation. The ongoing renovation and modernization programme for generating stations will be given added impetus from private sector participation.

Coal thermal power plants account for 70 per cent of the generating capacity in the country. This is likely to be the scenario in the future too. The efforts aimed at the efficient use of energy and efficient energy production thus are aimed at both optimizing the utilization of resources as well as minimizing the environmental pollution. A major emphasis in this direction is on promoting clean coal technologies and another is on promoting renewable energy sources.

India's renewable energy programme can be traced to the first oil shock in the early 1970s. Renewable energy technologies have since been developed and implemented in India for power generation, heating, cooling, lighting, pumping, cooking devices and as alternative fuels for surface transport. About 1050MW of power-generating capacity – 7 per cent of the total power capacity added during the Eighth Plan – was made up of non-conventional energy sources (GoI, 1997). A significant amount came from grid-connected wind farms, where capacity rose from 47MW in 1992 to 900MW in 1997. Even so, renewables comprise less than 2 per cent of the total installed capacity. The Ministry of Non-conventional Energy Sources has also announced certain targets to be achieved in the Ninth Plan, including (MNES, 1998):

1 3000MW of additional power from renewables;
2 The installation of 10,000 PV water-pumping systems;
3 Solar water-heating systems for 500,000 households;
4 500,000 m² of solar collectors in industrial/commercial establishments; and
5 Additional small hydro capacity of up to 15MW.

Potential CDM Options for India

As is true for most developing countries, the major potential for mitigation exists in the energy sector, both on the demand-side and supply-side. The 1990 GHG inventory for India shows the energy sector as the main emitter of CO_2, accounting for 87 per cent of the total emission, the rest being emitted from the cement industry (4 per cent) and emissions from land conversion (9 per cent). Fuel combustion in the industrial sector accounts for 41 per cent of total emissions in the energy sector and the power generation 34 per cent. The largest growth in emissions for the above period was in the power-generation sector, which grew at an annual rate of 8.6 per cent whereas emissions from the industrial sector grew at 4.7 per cent annually.

The industrial sector consumes about 50 per cent (including feedstock) of the total commercial energy produced in the country. The total industrial energy consumption, including non-energy uses, grew from 45.7MTOE in 1984–1985 to 113.1MTOE in 1996–97. The major energy-intensive industries are fertilizer, iron and steel, aluminium, cement, and pulp and paper. These industries account for almost 65 per cent of the total industrial energy consumption. Table 8.7 reports the estimates of energy consumption for these industries. Energy consumption for these industrial sectors was estimated based on energy consumption norms and existing production capacities. The current energy consumption norms are greater than international norms and substantial scope exists for upgrading technologies in these sectors. Very few estimates exist for emissions from individual industrial sectors. One such estimate for the steel industry shows an increase in emissions from 74.3Gg CO_2 in 1990 to 93.8Gg CO_2 (Shukla and Garg, 1999).

Table 8.7 *Estimated energy consumption in major industrial sectors (1990)*

Energy consumed in different sectors	G Kcal
1 Iron and steel sector	147,035
2 Fertilizer	149,533
3 Cement	68,899
4 Aluminium	32,588
5 Pulp and paper	20,600
Total energy consumed in the above sectors	418,655
Total energy consumed in the industrial sector	641,580[a]

a) Calculated from specific energy consumption and capacity tables in TERI, 1998b, pp237–252

Despite initiatives to promote clean energy, the expansion of conventional energy generation, particularly coal-based generation, implies a rapidly growing emissions profile for India. Hence, one of the major sources of emissions in the future will be power generation. According to the Fifteenth Electric Power Survey, the All-India peak demand is estimated to reach 176GW by the year 2012 and a total energy requirement of 1058 billion units (TERI, 2000). Coal-fired power generation will form a major share of new capacity additions. The projections indicate that the opening of the markets for the import of petroleum and gas will lead to some substitution, but a huge abundance of coal in the country will guarantee that it will still be the dominant source of energy. Under business as usual, CO_2 emissions in 2010 may be three times greater than 1990 levels in the energy sector (TERI, 1998d).[19] Table 8.8 also illustrates CO_2 emissions from the forestry sector and CH_4 from the agricultural sector. Although relatively small in magnitude, forestry and land-use emissions are expected to grow rapidly. Agricultural emissions of CH_4 account for 32 per cent of the total CO_2 equivalent emission, but are expected to grow at a slower rate.

Current changes to development paths, however, might prevent GHG emissions reaching such levels (Reid and Goldemberg, 1997). CDM finance would obviously strengthen the trend towards reduced emissions.

The inventory presented above reveals that the major increase in emissions over the next 20 years would be related to energy consumption in the economy. The projected energy emissions indicate that energy efficiency and the increasing use of renewable energy (or, more broadly, the move towards low carbon options) are the two main measures that can reduce emissions substantially. Power generation, and specifically coal-based power generation, has great potential for mitigation. Transmission and distribution losses are other major sources of energy losses and, hence, emissions. There is considerable scope for reducing losses and, hence, mitigation potential. Mitigation options in the power sector include clean coal technologies and renewables. Options such as bagasse-based cogeneration and combined cycle plants are already profitable and generate fewer emissions per kWh of electricity than conventional generation.

Table 8.8 *Projections of CO_2 and CH_4 emissions in the baseline case (Tg)*

		1990	2000	2010	2020
CO_2 emissions (Tg)	Energy sector	532	973	1555	2308
	Forestry/land-use sector	2	29	77	–
Total		534	1002	1632	(2308)[4]
CH_4 emissions	Enteric fermentation[1]	143	154	168	183
(in CO_2 GWP	Manure management[2]	19	21	22	23
equivalents, Tg)	Rice[3]	85	96	101	108
Total (CH_4 only, Tg)		247	270	292	314
Total (CO_2 and CH_4, Tg)		781	1272	1924	(2622)[4]

Notes:
1 The emission projections are made by the Central Leather Research Institute based on livestock growth rates for the period 1981–87
2 The emissions from manure management are also based on same animal population growth rate
3 National Physical Laboratory estimates
4 Totals exclude forestry/land-use sector
Source: TERI, 1998d

Alternative low carbon fuel options to current energy sources are a major mitigation option. Fuel switching to lower carbon fuels is an option, which to some extent will be taken up by the economy on its own. This, though, is a short-run option. In the medium and long run, renewable energy, both centralized as well as decentralized options, are important mitigation options. With a vast rural population and a fair share of remote areas, renewable energy could provide an avenue of providing clean energy and prosperity to these communities. Small hydro, wind and biomass-based power, although relatively more expensive than conventional coal-based plants, provide significant abatement opportunities. Renewable options for irrigation in the agricultural sector provide a reliable energy option.

Industrial process energy consumption is another area where substantial reductions can be achieved. These include both efficient processes as well as other demand-side management options. As was highlighted in the preceding section, the industrial sector accounted for 41 per cent of emissions during 1990–91 in the economy. Typically, these options lead to an improvement in energy efficiency and resource conservation, and introduce advanced technologies, so laying the foundation for long-term sustainable development. Opportunities in the iron and steel and cement sectors are significant. The NFYP too lays emphasis on it and states that the 'adoption of energy-efficient technologies in major energy-intensive industries like iron, steel, chemicals, petroleum, pulp paper and cement' are one of the objectives for the energy sector.

India already has five CDM-type projects in place under the pilot version of the CDM, Activities Implemented Jointly (see Table 8.9). The projects cover the spectrum of potential CDM activities including improvements in industrial efficiency, power generation using waste materials and changes in agricultural practices. Such diversity is consistent with the use of Activities Implemented Jointly as a learning period. To some extent these activities also reflect the priorities of the government. Further insight into government priorities is provided by the overview of mitigation options in India funded by the Global Environmental Facility, which is given in Table 8.10.

In this chapter, we review 22 separate projects in five different sectors (see Table 8.11). These options form the basis of a comparative ranking using analytical techniques. These options were selected on the basis of three main criteria: their consistency with national development priorities; the relatively high level of energy consumption in the base activity; and the relatively large GHG reduction potential offered by the abatement technology.

Table 8.9 *Pilot phase – Activities Implemented Jointly projects endorsed by the Government of India*

Name of project	Location	Investing party	Host party
1 Direct reduced iron	Gujarat	Japan	M/s Essar, Gujarat, India
2 Energy recovery from waste gas and liquid	Gujarat	Japan	IPCL Plant at Vadodara, Gujarat, India
3 Integrated agricultural demand-side management	Andhra Pradesh	World Bank	Andhra Pradesh State Electricity Board, India
4 DESI Power: biomass	At 20 sites	The Netherlands	DESI Power, Development Alternatives, India
5 Tamarind orchard agro-forestry for dry land	Karnataka	USA	ADAT, Bagepalli, Karnataka, India

Conventional Power Generation

Improvement in coal-based power generation is certainly possible. The average gross conversion efficiency of coal thermal power stations in India is 28 per cent and the average net efficiency is about 25 per cent (CEA, 1990). A variety of new technologies are available that would result in higher energy efficiency from conventional fuel sources, mostly coal. In addition to reducing CO_2 emissions per unit of output, these technologies lead to significant reductions in, or even elimination of, particulates, SO_2 and NO_x releases. The introduction of such technologies would also constitute 'leap-frogging' for the Indian power sector, allowing them to benefit directly from technologies that have been developed elsewhere. Over time, a growing familiarity with such technologies may facilitate further transfer and even result in domestic production.

Two technologies imply fuel switching away from coal, to natural gas and bagasse respectively. The first option uses natural gas in a combined cycle technology to attain efficiency levels of 50 per cent or more – higher than any coal option. Co-benefits would be commensurately higher than for coal options, too. The bagasse cogeneration option entails sugar mills developing a capacity to burn bagasse produced as a by-product from sugar-refining. This produces diverse social benefits, ranging from additional income for farmers and sugar-mill owners to increased employment opportunities. Even without a credit for carbon reductions, this option is profitable, but could be greatly extended with further financing.

Table 8.10 *Mitigation initiatives funded by the Global Environmental Facility (GEF) in India*

Projects	GEF allocation in US$ million		
Implementing agencies	*UNDP*	*World Bank*	*IFC[a]*
Operational projects			
• Optimizing development of small hydel resources in hilly regions of India	7.500		
• Development of high-rate biomethanation process as means of reducing GHG emissions	5.500		
• The ALGAS project	0.850		
• Alternate energy development project		26	
• Selected options for stabilizing GHG emissions	1.500		
• Coal-bed methane capture and utlization	9.900		
Project preparation grants			
• Carbon-emission reduction through biomass energy	0.196		
• Fuel-cell buses			
Pipeline projects			
• India energy efficiency project		5	
• Solar thermal electric project		49	
• Photovoltaic market transformation initiative – regional project		–	15

a) International Finance Cooperation

Table 8.11 *Potential CDM projects*

Sector	Base case technology	Mitigation options
Conventional power generation	Pulverized coal subcritical boilers for all cases	Pulverized coal super-critical boilers Atmospheric fluidized bed combustion Pressurized fluidized bed combustion Integrated gasification combined cycle Combined cycle gas turbines – natural gas-based Cogeneration – bagasse-based
Renewables for power generation	Traditional thermal power generation using domestic coal	Small hydro Wind farm Photovoltaic-based power generation Biomass-based power generation
Renewables for agriculture	Small-scale diesel pump-sets	Agro-waste-based gasifier Wood-waste-based gasifier Wind shallow pump Wind deep pump PV pump
Iron and steel	Integrated steel plant with wet-quenching option Wet quenching Open hearth furnace Conventional electric arc furnace Conventional (normal) casting	Direct reduction process Dry quenching Basic oxygen furnace Ultra high-powered electric arc furnace Continuous casting
Cement	Non-precalciner kilns No preheater kilns	Dry precalciner kilns Dry suspension preheater kilns

Renewable Energy Sources and Technologies

India has one of the oldest renewable energy programmes in the world. The primary mover initially was the first oil shock, and development of the renewable energy programmes was thought of as an energy security measure. Over the years a number of other compelling reasons have added to the drive for increased use of renewables. Some of these are:

- The inability of conventional systems to meet the growing energy demands in an equitable and sustainable manner;
- The large-scale and negative impact of conventional energy production and consumption on the physical and human environment;
- The need to meet the energy needs of the unserved population in rural and remote areas, as well as those residing at islands; and
- The need for maintaining a properly diversified energy mix.

Renewable options were considered for both power generation, and for power supply and irrigation in the agricultural sector. In the former, four options – small hydro, wind farms, biomass gasifiers and photovoltaics – are examined. Decentralized power may enhance local energy supply and create job openings either in maintaining power generation (for example, through energy plantations for biomass), or because increased power supply facilitates economic growth. Greater demand for biomass gasifiers and photovoltaics would also create job opportunities in the manufacturing sector. Adverse environmental impacts are minimized or non-existent. Similar benefits hold for the agricultural sector. Estimates of the potential for renewables are given in Table 8.12.

Table 8.12 *Estimates of technical potential for renewable energy technologies*

Source/systems	Approximate potential
Biogas plants (number)	12 million
Improved cook-stoves (number)	120 million
Biomass energy	17000MW
Solar energy (thermal and electrical)	35MW/sq km
Wind energy	20,000MW
Small hydro power	15,000MW
Ocean energy	79,000MW

Iron and Steel

The iron and steel sector in India has grown substantially in recent years. Total production of finished steel was 17 million tonnes in 1994–1995, with domestic demand projected to rise to 31 million tonnes by 2001–2002.[20] Such high growth is driven by large-scale investment in the infrastructure, growth in the transportation sector and the increasing demand for consumer durables. Low per capita consumption of only 22kg – among the lowest in the world – leaves significant room for growth (TERI, 1998a). Eight million tonnes of new capacity is currently being developed by the private sector.

The iron and steel industry is the largest industrial consumer of energy, with energy costs making up 30 per cent of production expenditure. However, energy consumption per unit of output is high in comparison to many other countries – approximately twice the level of OECD producers and higher than the main developing country competitors (TERI, 1998a). This is for several reasons. Low-grade ores used by the Indian industry require more energy per unit of output. Most plants in India still use the dated wet-quenching method for coke preparation. Modern dry-coke quenching improves the coke quality and reduces energy consumption and dust emissions. Efforts are being made to reduce energy consumption by importing good quality coal for cooking, developing heat energy-recovery systems, phasing out the open hearth furnaces for steel-making and a increasing continuous casting capacity in industry.

Potential CDM activities for the iron and steel sector are detailed in Table 8.11 (see p276). These are based on introducing current technology into all or part of the steel-making process. Such technology will lower CO_2 emissions and simultaneously raise the efficiency of steel-making in India. Unfortunately, lack of adequate data prevents estimation of carbon price in terms of US$ per tonne of carbon, so improvements are expressed only in terms of reductions per tonne of finished steel.

Cement

India is the world's fourth largest producer of cement after China, Japan and the US. The cement industry is one of the six most energy-intensive industries in India (TERI, 1998a). Cement is manufactured in India using dry, semi-dry or wet processes, with current trends favouring the dry manufacturing process. This is mainly due to the relatively lower thermal energy consumption of the dry process in comparison to the wet process, even though it requires greater electricity consumption (CMA, 1996). The cement industry is a continuous process industry where power cuts not only disrupt production but can entail costly damage to equipment. In order to maintain uninterrupted cement production, the use of captive power by the cement industry is increasing – 27 per cent of the total power supply is currently from captive power, compared to 16 per cent at the start of the decade (CMA, 1997).

Two particular abatement options are reviewed here: the introduction of dry precalciner kilns and dry suspension preheater kilns. The first technology allows 85–90 per cent of the necessary pretreatment, or 'calcination', of the 'meal' input, to take place at a lower temperature than that used in the main kiln. As a consequence, the energy consumption of the calcination process is significantly reduced relative to present technologies which rely on an unnecessarily high temperature from the outset. The second technology makes use of waste heat recovered from the main kiln for use in a preheater kiln.

Prioritizing CDM Options

Which of the above options is best depends on one's perspective. Investors from developed countries will be concerned primarily with a project's abatement cost, financial risk and feasibility, while India will be more concerned with the development and environmental benefits that arise, and their consistency with national priorities. Different parties to a transaction will weigh project characteristics differently, suggesting a priori that participants from developed and developing countries would rank overall projects differently.

To explore how projects meet national priorities and which, if any, would satisfy all parties, projects were assessed both on the basis of their development benefits and their carbon abatement cost. Projects were evaluated against three main criteria, which were further subdivided into three subcriteria each, as listed in Table 8.13.[21] The criteria were based on the development perspective set by the NFYP and are as follows:

- Incremental cost-effectiveness: the incremental cost per unit of CO_2 reduction;
- Feasibility: including (1) consistency with existing government policy; (2) perceived risks from both the investors and the host country's point of view; and (3) consistency with national priorities – ie, social and economic development;
- Other environmental benefits: including (1) resource conservation; (2) pollution loading; and (3) health impacts; and
- Development: including (1) the importance of employment generation; (2) adding value; and (3) rural development.

The ranking of projects is difficult because many important criteria are non-quantifiable and non-comparable. It is not clear, for example, how a project that leads to improvements in air quality should be compared with a project that leads to greater employment opportunities. To tackle this problem, an analytical hierarchical process (AHP) technique (Saaty, 1980) is used. This provides a tool for the scoring and weighting of non-quantifiable attributes of a mitigation option. While not providing single answers, the AHP results reveal which projects will be preferred under different preferences and how rankings would change if certain criteria were to be given more weight. It lends transparency and structure to project evaluation and decision-making.

To assess and prioritize the mitigation options the HIPRE 3+ (hierarchical preference analysis) model developed by the Systems Analysis Laboratory of the Helsinki University of Technology has been used. The process involves ranking the alternatives – in this case the GHG mitigation options – by specific evaluation criteria, described above. Since these criteria and subcriteria are qualitative in nature, the CDM project alternatives are evaluated against each other, for each subcriteria, using a pair-wise comparison approach. Similarly, the criteria are evaluated using a pair-wise comparison. Thus, for example, any two mitigation alternatives for power generation – say, integrated gasification combined cycle technology or pulverized coal super-critical boilers – can be compared for their relative importance on the subcriteria of, say, resources conservation. Similarly, the subcriteria of resource conservation could be compared with the subcriteria of health for assessing their relative importance in the criteria called 'other environmental benefits'. Each pair being compared is scored according to the following:

1   A is equally preferred to B.
3   A is slightly preferred to B.
5   A is more preferred to B.
7   A is significantly preferred to B.
9   A is strongly preferred to B.

Once such comparisons are made and scored for each definable pair, the AHP model arrives at a composite weighting of criteria and the ranking of projects. The model also allows the ranking of projects by specific criteria instead of a composite ranking.

In the Indian case, the views of senior government officials and leading experts were sought so as to arrive at a prioritized list of projects that would reflect national development priorities while mitigating emissions of GHGs. Since these officials/experts had difficulties in arriving at consistent pair-wise choices, they were asked to weight the relative importance of the subcriteria and criteria defined in Table 8.13. This table also provides a summary of the relative weights as derived from the survey. Each project was then scored against these weighted criteria to arrive at a project ranking reflecting national priorities. The prioritization exercise was applied separately to all the sectors reviewed. Projects were ranked both by carbon abatement cost alone, and by a measure of overall development benefits. Table 8.14 illustrates how the two rankings compare.

Interestingly, for three of the four sectors for which comparisons can be made, the two highest ranked options, based on the cost of carbon offset, are also the two highest ranked options based on their co-benefits. This suggests that there is a high degree of overlap between those projects that would be prioritized by carbon-focused investors and those that are in the best interest of India. Only the renewables for power generation category shows a mismatch in the first choice as perceived by market criteria alone and the wider criteria that make up the AHP scores. Also, in conventional power generation and renewables for agriculture, some of the lower rankings are different.

Conclusions for the India Case Study

For the abatement options reviewed, there seems to be a high degree of overlap between those projects that are available at the lowest cost and those which are

Table 8.13 *Base-case weights for the various criteria in the AHP exercise*

Main criteria	Relative weights (as percentage of total weighting)	Subcriteria	Relative weights (as percentage contribution to the main category)
Incremental cost-effectiveness (US$/tC)	7.5		
Feasibility	39.4	Government policy Compromising socioeconomic development Risk	42.9 42.9 14.3
Other environmental benefits (non-CO_2)	13.7	Resource conservation Decrease in pollution loading Health	42.9 14.3 42.9
Development	39.4	Employment generation Value addition Rural development	25.8 10.5 63.7

Table 8.14 *Summary of sector rankings and carbon price*

	Carbon price (US$/tonne of carbon)	Ranking by carbon price	Ranking by AHP approach
Conventional power generation			
Bagasse-based cogeneration	−244.1	1	1
Combined cycle generation (natural gas)	−133	2	2
Atmospheric fluidized bed combustion	7	3	5
Pressurised fluidized bed combustion	47	4	4
Pulverized coal super-critical boilers	96	5	6
Integrated gasification combined cycle	96	5	3
Renewables for power generation			
Small hydro	29	1	2
Biomass power	134	2	1
Wind farm	216	3	3
Photovoltaic	1306	4	4
Renewables for agriculture			
Wood-waste-based gasifiers	169	1	1
Agro-waste-based gasifiers	177	2	2
Wind shallow well	298	3	5
Wind deep well	329	4	4
PV pump	6333	5	3
Cement			
Dry suspension preheater kilns	7	1	1
Dry precalciner kilns	214	2	2

most consistent with national priorities and offer most to India in terms of development and environment co-benefits. The options considered here are also among the ones that have been suggested by most governments to be included in the CDM of the Kyoto Protocol. Renewable options and energy-efficiency improvement are the common minimum set that is acceptable to most. There is a considerable debate, though, on the coal technologies, the logic being that by promoting coal projects in CDM, the long-term trajectory is tied to carbon-intensive fuels. From the Indian standpoint, coal is the cheapest and the most secure source of fuel. This implies that coal will form the mainstay of energy requirement for a considerable time to come. CDM provides an avenue for moving this high emission pathway to a lower emission pathway.

There are two improvements that could be made in the analysis. One would be to extend the analysis to compare all projects together, rather than merely within sectors. This would allow policy-makers to know which sectors should be emphasized, before focusing in more detail on projects within a sector. This will require a weighing of each sector to project the relative importance of the individual sectors in relation to overall economic growth. From the investors' standpoint, the total potential in each sector and the transactions costs are important factors.

The study does not include two major sectors – namely, transportation and carbon sinks. The sinks have been deliberately excluded, given the firm Indian stand on not including sinks in CDM. Given the present policy framework and management structure where the private ownership of forests is completely unacceptable, the possibility of allowing sinks projects, even if it is agreed at the international level, is very unlikely. The transport sector is another potential area for financing. Urban air pollution and congestion have been identified as key environmental problems and are the direct consequence of numerous inefficient motor vehicles. The CDM could be used to help finance new transport initiatives or to improve engine technologies. Finally, more details on the projects themselves would allow for more accurate evaluation. For example, the health benefits of cleaner coal technologies are sensitive to the precise location of the plant under consideration.

NOTES

1   This study of Brazil is based on a background report presented to the World Resources Institute in which the results of economic and social viability were first presented in Serôa da Motta, Ferraz and Young (2000) and summarized in Austin et al (2000).
2   See, for example, a survey in Ekins (1996).
3   See, for example, Brown, Cabale and Livernesh (1997) and Adger and Brown (1994).
4   See, for example, Fearnside (1995).
5   For a survey, see Serôa da Motta (1997).
6   It was 50 per cent until two years ago.
7   See, for example, Amaral et al (1998).
8   See Almeida and Uhl (1995).
9   We should note that the above does not allow for any future increases in the price of land. If they were to be built in, this would reduce the return to sustainable logging.
10   Energy conservation is already well advanced in Brazil and costs associated with its expansion varies a lot; solar energy is still very expensive. We excluded these two important options, therefore, from our analysis, although they must be recognized as such.
11   The latter is an intermediate value between 230KWh/t (Zylberstajn and Coelho, 1992) and 350KWh/t (Tecnorte/Fenorte 1995, quoted in Freitas, Caetanoe and Cecchi, 1997).
12   See IE (1998).
13   An average of carbon benefits for plantations for pulp, charcoal and sawlog, considering their current production share on total output from planted forests in Brazil.
14   This practice has been forbidden near urban centres in the state of São Paulo, the largest producer.
15   See Serôa da Motta and Mendes (1996).
16   Asia Least-Cost Greenhouse Gas Abatement Strategy (ALGAS) was a GEF-enabling activity project funded through the Asian Development Bank. TERI undertook mitigation options analysis of this project for India.
17   The Indian development perspective is enunciated through the Five Year Plans formulated by the Planning Commission.
18   Energy GDP elasticity is defined as the percentage change in energy consumption for every per cent change in GDP.
19   These are CO_2 emissions from the energy sector in a baseline scenario for the period 1990–2020, assuming a 12 per cent discount rate. They were calculated using the MARKAL model.
20   This is based on moderate estimates of growth of GDP (5–6 per cent per annum).

21 In the formal analysis and rankings shown later in Table 8.13, some recognition was also given to the carbon abatement cost. However, this amounted to only 7.5 per cent of total input to the project ranking and did not alter the overall rankings.

REFERENCES

ADB (1994) *Climate Change in Asia*, Asian Development Bank, Manila, The Phillipines

Adger, N and Brown, K (1994) *Trees, people, the missing link and the greenhouse effect*, CSERGE Working Paper, CEC 94–14, London

Almeida, O T and Uhl, C (1995) 'Identificando os custos de usos alternativos do solo para o planejamento municipal da Amazônia: o caso de Paragominas (PA)', in P May (ed) *Economia Ecológica*, Ed Campus, Rio de Janeiro

Amaral, P, Veríssimo, A, Barreto, P and Vidal, E (1998) *Floresta para Sempre: Um Manual para Produção de Madeira na Amazônia*, WWF/IMAZON/USAID, Belém

Austin, D, Faeth, P, Serôa da Motta, R, Ferraz, C, Young, C E F, Ji, Z, Junfeng, L, Pathak, M, Shrivastava, L and Sharma, S (2000) *How Much Sustainable Development Can We Expect From The Clean Development Mechanism?*, World Resources Institute, Washington, DC

Brown, P, Cabale, C and Livernash, R (1997) *Carbon Counts: Estimating Climate Change Mitigation in Forestry Projects*, World Resources Institute, Washington, DC

CEA (1990) *Public Electric Supply, All India Statistics, General Review 1986–87*, Central Electricity Authority, Government of India, New Delhi

Chomitz, K M and Kumari, K (1996) *The Domestic Benefits of Tropical Forests: A Critical Review*, CSERGE Working Paper, GEC 96–19, London

CMA (1996) *Indian Cement Industry (Statistics) 1995–96*, Cement Manufacturers' Association, New Delhi

CMA (1997) *36th Annual Report 1996–97*, Cement Manufacturers' Association, New Delhi

COPPE (1998) *Relatório das emissões de carbono derivadas do sistema energético – abordagem top-down* (http://www.mct.gov.br/gabin/cpmg/climate/programa/port)

CPCB (1990) *Status of Water Supply and Waste Water Collection, Treatment and Disposal in Class I Cities – 1988*, Central Pollution Control Board, New Delhi

Ekins, P (1996) 'The secondary benefits of CO_2 abatement: how much emission reduction do they justify?', *Ecological Economics*, Vol 16, No 1

EPTRI (1995) *Status of Solid Waste Disposal in Metropolis*, Environment Protection Training and Research Institute, Hyderabad (draft)

Fearnside, P M (1995) 'Global warming response options in Brazil's forest sector: comparison of project-level costs and benefits', *Biomass and Bioenergy*, Vol 8, No 5, pp309–322

Freitas, M A V, Caetanoe, M M and Cecchi, J C (1997) *Perspectivas da produção e consumo de álcool carburante no Estado do Rio de Janeiro*, COPPE, Rio de Janeiro

Funatura (1992) *Cost of Implantation of Conservation Units in Legal Amazonia*, Funatura/SCT-PR/PNUB, Brasília

GoI (1997) *Annual Plan 1996/97*, Planning Commission, Government of India, New Delhi

Heintz, R J and Tol, R S (1996) *Secondary benefits of climate control policies: implications for the Global Environmental Facility*, CSERGE Working Paper, GEC 96–17

IEA (1998) *Key World Energy Statistics form the IEA*, International Energy Agency, Paris, Cedex 15

IE-Instituto de Economia, UFRJ (1998) *Energia e desenvolvimento sustentável*, IE/UFRJ, Rio de Janeiro

IPCC (1996) *Climate Change 1995*, Cambridge University Press, Cambridge

MNES (1998) *Annual Report 1998*, Ministry of Non-conventional Energy Sources, New Delhi

Ravindranath, N and Somashekar, B S (1995) 'Potential and Economics of Forestry Options for Carbon Sequestration in India', *Biomass and Bioenergy*, Vol 8, No 5, pp323–336

Reid, W and Goldemberg J (1997) *Are Developing Countries Already Doing as much as Industrialized Countries to Slow Climate Change?*, WRI Climate Note, World Resources Institute, Washington, DC

Saaty T L (1980) *The Analytical Hierarchical Process,* McGraw Hill, New York

Serôa da Motta and Mendes, F (1996) 'Health costs associated with air pollution in Brazil', in P May, and R Seroa da Motta, *Pricing the Planet,* Columbia University Press, New York

Serôa da Motta, R (1997) 'The economics of biodiversitiy: the case of forest conversion', in *Investing in Biological Diversity: The Cairns Conference,* OECD, Paris

Serôa da Motta, R, Ferraz, C, Young, C E F (2000) 'Brazil: CDM opportunities and benefits', in, D Austin, and P Faeth (eds) *Financing Sustainable Development with the Clean Development Mechanism,* World Resources Institute, Washington, DC

Shukla, P R and Garg, A (1999) *GHG emissions inventory for India: sectoral and regional analysis* (www.climatechangesolutions.com/climate/ghgeifi.htm)

Tecnorte/Fenorte (1995) 'Parque de alta tecnologia do Norte Fluminense – sumário de subprojetos de empresas de base tecnológica – EBT. Campos', mimeo

TERI (2000) *TERI Energy Data Directory and Yearbook (TEDDY) 2000/2001,* TERI, New Delhi

TERI (1998a) *Activities Implemented Jointly in India,* TERI Project Report No 97GW54, pp100, TERI, New Delhi

TERI (1998b) *TEDDY 1998/99,* TERI, New Delhi (draft)

TERI (1998c) *Looking Back to Think Ahead: GREEN India 2047,* TERI, New Delhi

TERI (1998d) *India National Report on Asia Least-cost Greenhouse Gas Abatement Strategy (ALGAS),* TERI Project Report No 95GW52 (draft): June, TERI, New Delhi

UNEP (1997) *Asia Pacific Environment Outlook,* Bangkok, Thailand: United Nations Environment Program, Environment Assessment Program for Asia and the Pacific, World Bank, South Asia Regional Office, Washington, DC

UNFCCC (1997) 'Kyoto Protocol to the United Nations Framework Convention on Climate Change (UNFCCC)', FCCC/CP/1997/L.7/Add.1, Bonn

Zylberstajn, D and Coelho, S T (1992) 'Potencial de geração de energia elétrica nas usinas de açúcar e álcool brasileira, através da gaseificação da cana e emprego de turbina a gás', *Revista Brasileira de Energia,* Vol 2, No 2, pp53–72

Index

Page numbers in *italics* refer to boxes, figures and tables